THE MEN WHO BREACHED THE DAMS

In preparation by the same author:

Beyond the Dams to the Tirpitz
617 Squadron

The Men Who Breached the Dams

617 SQUADRON
'The Dambusters'

Alan W. Cooper

WILLIAM KIMBER · LONDON

First published in 1982 by
WILLIAM KIMBER & CO. LIMITED
Godolphin House, 22a Queen Anne's Gate,
London, SW1H 9AE

© Alan W. Cooper, 1982
ISBN 0-7183-0308-3

Reprinted 1983

Photoset in North Wales by
Derek Doyle & Associates, Mold, Clwyd
Printed in Great Britain by
Redwood Burn Limited, Trowbridge

Contents

List of Illustrations

Maps and diagrams

Acknowledgements

The research for this book is based on the records and recollections of a great many people, and was made possible by their co-operation and that of many institutions. A list of those to whom I am indebted can be found at the end of this book, but in particular my thanks are due to my friend Horst Muller, W/C Wally Dunn, John Evans and to Norman Franks for giving me the benefit of his experience and expertise in putting this book together.

My thanks also are due to the Imperial War Museum, Public Records Office, Royal Air Force Museum, Air Historical Branch, Commonwealth War Graves, Royal Air Force Records Gloucester and Adastral House, and last but not least the staff of William Kimber & Co.

Foreword

by

Air Marshal Sir Harold Martin KCB, DSO, DFC, AFC, RAF, Rtd.

When the British thirty year Official Secrecy Regulation was lifted in 1974/75 for papers selected to go to the official archives for 1944/45, the interest of Alan Cooper who is an official at the Public Records Office was caught by the story of the RAF attack on the Ruhr dams on the night of 16/17 May 1943. Over the last eight years he has carefully gathered all information he could acquire from the records made available, plus accounts from people, mainly industrial and RAF ground personnel, who in one way or another had been connected with the mounting of the raid. He has also talked to a number of the few surviving aircrew. He now presents this mass of data in book form with emphasis on the men of the actual 'Dambuster Squadron' with a highly commendable result.

Inevitably he found inconsistencies which cannot always be reconciled. Memories fade and personal accounts vary. Preserved records are by no means complete as space was not available to store all documents; recording in any event was frequently regarded as a tiresome bore and sometimes dashed off. Despite this Alan Cooper has endeavoured to sort out as true a story as is possible. In particular the records of the Germans emerge as confused, as well they might, having been taken completely by surprise by ultra low flying aircraft in the cold hours after midnight. From monumental wreckage they tried to piece together what had struck them. Some reports were made five months after the event and submitted to higher authority eight months later, nevertheless they are interesting if only to demonstrate the complexity and confusion of wartime reporting and communications.

I personally am impressed by the graphic accounts from the

surviving German gunners who were deployed on the Möhne Dam. It must have been an awesome experience suddenly to face huge bombers racing across the dam wall at sixty feet with all guns firing, followed by a menacing weapon bouncing independently behind the aircraft from a release point over four hundred yards back, with splashes going up at each bounce: then great spouts of water lifting many hundreds of feet and falling down like pouring rain and large waves lapping over the dam wall as the weapon exploded. I remember the German performance well and recall with admiration the time when their guns at long last, either through blast damage or shortage of ammunition, fell silent.

Alan Cooper has adhered to the general account by the late Wing Commander Guy Gibson VC, DSO, DFC, as recorded in his own book *Enemy Coast Ahead*. This is right as Gibson was the raid leader who fully understood the operation from almost the beginning to the end. After the experimental drops of the Wallis bouncing bomb by a Vickers test pilot, Gibson took part in the final operation trials. He oversaw the special low level training of the aircrews, personally led the attacks on the Möhne and Eder dams, saw each attack go in and reported from the spot. He was a matchless leader and airman and no VC could have been more deserved.

The extraordinary genius of Barnes Wallis and the imaginative skill of other scientists and engineers are the very foundation of the story. Wallis was brilliant yet perhaps the strongest element of his character was compassion. It befell my lot to seat him on his bed about 08.30, distraught at the operational loss of 8 out of 19 aircraft. He kept repeating, 'I would never have gone ahead had I known the cost.'

I like very much the importance given to all 617 Squadron members who did not fly but played a vital part by providing the engineering and administrative support without which an operation of this weight could never have taken place.

I would like especially to praise the work and care Alan Cooper has devoted to following faithfully to the end the fate of those of us who did not return.

'Mick' Martin

CHAPTER ONE

The Idea

The concept of an attack on the Germans dams in the event of a second world war was first discussed by the Air Ministry's Bombing Committee in 1938. This committee had been formed in the 1930's to study and assess how Great Britain could hit, damage or destroy vital German targets, targets whose loss would have a severe effect on Germany's ability to wage war. Many targets were suggested, listed, discussed and agreed upon. Many of them could be attacked by conventional bombing; others, of a more difficult nature, would need specialised treatment.

One of the latter, discussed in committee on 26th July 1938, was the dams situated in South Westphalia. The objectives were clear:

(a) Cut off essential supplies of water for industrial and domestic purposes.
(b) Cause flooding and damage to industrial plants, railways, waterways, etc, in the river valley.
(c) And/or to prevent the maintenance of sufficient water for navigation in the inland waterways system.

Only eleven days earlier, on 15th July, the economic and strategic importance of these dams was discussed by the Plans (Op) Operations at Air Ministry. The dams were:

(1) The Möhne – situated in the Möhne Valley south-east of Dortmund, whose role was to collect rainfall to prevent winter flooding and to provide power for electrical generators. Of utmost importance was the part it played in sustaining the underground water supply vital for industrial and household supplies.
(2) The Eder – situated south of Kassel and south-east of the

Möhne. It was built to act as a reservoir for the important Mittelland Canal that runs from the Ruhr to Berlin. It also prevented flooding of farmland in winter and finally served hydro-electrical power stations.

(3) The Sorpe, Ennepe and Lister dams – situated south of Dortmund and south-west of the Möhne. The roles of these dams were similar to that of the Möhne.

In total there were seven dams in South Westphalia, but the Möhne, Eder, Sorpe and Ennepe were considered of prime importance. The destruction of the one outstandingly important Möhne dam alone with the massive loss of hydro-electric power, would have serious repercussions on the production output from the Ruhr, Germany's major industrial area.

The most favourable time for an attack upon these dams would be after a period of heavy rain when the reservoirs were full. Nevertheless, although flooding would not be so severe if the dams were breached during a dry period, a serious and immediate shortage of water would result. However, it had to be borne in mind that the destruction of the dams would be more difficult at low water. Yet, if the Möhne dam, holding some 130 million cubic metres of water, was breached, the force of water flowing down in the Ruhr Valley in a few short hours would be so powerful that villages and towns as well as waterways in the Ruhr Valley, as far as the Ruhr itself, would be swept away and destroyed. The entire area's population of between four to five million would be without water, and the mines and coke plants paralysed owing to a lack of industrial water supply.

However, although the dams presented a valuable target, a method of destroying them was far from straightforward. Normal bombing would hardly chip the huge structures and if a torpedo-like weapon were developed, surely the Germans would protect the dam walls by using anti-torpedo netting. So, the dams, recognised as important targets, remained on the Air Ministry's list; perhaps one day a method of destroying them might come to light.

Soon after the war began, a method was put forward, but it was viewed with a good deal of scepticism. The Controller of Armament

Research and Development at the British firm of Messrs. Vickers Armstrong Ltd, Weybridge, was Barnes Wallis. His brilliant mind had devised an idea.

Barnes Wallis, who celebrated his 52nd birthday in September 1939, was no stranger to the world of aviation design. He had been a design engineer during the First World War, had invented the Geodetic method of construction used in the Wellesley bomber (which he had designed in 1935) and later in the famous Wellington bomber.

Wallis's active mind knew only too well the importance of the dams to the German war machine, and had been toying with an idea for successfully attacking them, feeling that it was more than just possible. In the late autumn of 1939, a City banker entertained Group Captain Frederick Winterbotham in London, a member of the Air Staff at the Air Ministry. During their meeting Winterbotham was asked if he would like to meet Barnes Wallis, the inventor of the Geodetic system. Winterbotham said he would and a meeting was arranged at the Surrey home of Wallis in Effingham. Winterbotham recalls a subsequent meeting a year later:

At this time in the war the Germans were dropping 500-pound bombs on London which were causing great destruction out of all proportion to their blast effect, especially since they usually exploded deep inside a building. Wallis was determined to find out why this was happening. He soon had the answer; it lay in the anti-submarine depth-charge, detonated under water. It destroyed by means of shock waves and not by blast which was transmitted by the water itself. Barnes was making endless studies of the effect of shock waves, working out the possibilities with regard to size of the bomb, the depth of penetration and the probable results of the explosion.

Wallis had spoken earlier to Winterbotham of his ideas about the dams and the Group Captain acknowledged that there was something in what Wallis was trying to develop. However, what was needed was to have his plan adopted by the Air Staff for support to make a penetrating bomb. Winterbotham was able to by-pass orthodox methods of approach and wrote to a friend, Desmond Morton, who was a member of the Prime Minister's staff. Winterbotham also knew that if interested, Morton would seek a reaction from Professor Lindemann (later Baron Cherwell), who, like Morton, was a Personal Assistant to the PM. Morton's reply came on 5th July 1940:

> I have not only read your interesting paper on the ultimate aim of bombing warfare but have consulted certain goodwill experts without disclosing your identity. The view held is that such a project as you describe could not come to fruition until 1942, even if then.

Winterbotham passed the message to Wallis who was very disappointed but the Group Captain persuaded him to continue his research on the subject. This he did, and continued to press his ideas on anyone who would listen.

Eventually he persuaded the Ministry of Aircraft Production that the possibility of destroying a large dam was at least worth investigating. The Road Research Laboratory, under the direction of Dr William Glanville (later Sir William, who died in 1976 aged seventy-five) had been actively engaged in research on military problems from the outbreak of war and had considerable experience in using models for forecasting the effectives of explosives. Wallis discussed his problem with Glanville who immediately undertook to make a number of experiments using various models. The basic rule of model testing was comparatively simple. A scale model built of the same material as the original would be damaged in much the same way and to the same extent, if the weight of the explosives were produced by the cube of the scale ratio.

As both the Möhne and the Eder dams were gravity types, a significant question-mark stood against the validity of any model test, which had to be answered before firm forecasts could be made.

At this time the idea of a bouncing bomb had not been born, and the tests proceeded on the basis that a large, but more-or-less conventional bomb would be dropped from an operational altitude. As the Möhne was slightly more solidly built than the Eder and had a sealing bank of clay at its base on the upstream face, it was therefore chosen as the prototype for reproduction ratio of 1:50. On this scale the model, built by Doctors Davey and A.J. Newman, and which can still be seen by visitors to the Building Research Establishment, was of a practical size and the 6,800 kg. of explosive on the full scale, became the reasonable weight of 56 g. The first model was made as accurately as possible with scaled-down cement mortar blocks laid in cement, representing the cyclopean masonry of the real dam.

The first test was made with the 56 g. charge at a distance of 9 metres – about 29 feet, equivalent on the full scale of 45 metres from the upstream face – about 165 feet. (These measurement were supplied to the author by Dr A. Collins who was one of the engineers working directly on the project.) The result of the test was a vertical crack at the centre of the dam and a horizontal crack under the crest.

Further models were required, but making them with the mortar blocks was too time-consuming, so the RRL developed a sliding template to form layers of mortar representing masonry which cut model-building time to about two weeks. Two of the new models were tested, one with a clay foundation and the other with a more massive concrete base. The simplified models were found to be slightly weaker than the accurate model, but were judged more representative of the strength of the actual dams which would certainly contain some cracks through age.

For the Möhne and Eder dams, the charge required to produce an effective breach was estimated to be in the region of 3,600 kg, exploded 9 metres below water. This was within the capability of the aircraft that would be available, but it was still essential to obtain a direct hit. As there was no means of obtaining a direct hit with current aiming methods, the research could be said to have been successful but unexploitable! However, Barnes Wallis was not to be defeated and was already secretly evolving a plan.

Wallis telephoned Group Captain Winterbotham with the idea of a spherical bomb, detonated so that the explosion would reach all points of the surface at precisely the same moment. Winterbotham asked if a round bomb would penetrate deep enough to do any real damage with a shock wave. Wallis was keen to learn if there were any data at the Air Ministry on the subject and the Group Captain rang the Air Ministry to ask what the effect of dropping a large spherical bomb from about ten thousand feet would be. The reply was that it would bounce along like a football but with no accuracy at all.

Wallis was told of this answer. There was a pause then he said, 'Splendid'.

Wallis later telephoned Winterbotham again to ask if he could have a set of the drawings of the new Avro Lancaster bomber that was due to come into service shortly. The Lancaster was a four-engined version of the twin-engined Avro Manchester, which had proved a failure due mainly to lack of power. Wallis went on to say that he had spent all of one Sunday on the terrace of his home, shooting a glass marble at the surface of the water in a tin bath, with the help of his children, Mary and Christopher. How about, he suggested to Winterbotham, bouncing a bomb along the surface of the water against the dams?

Sufficient interest had now been generated for the Ministry of Aircraft Production to form an 'Air Attack on the Dams' Committee.[1] Its Chairman was Dr David Pye, Director of Scientific Research at MAP, and other members included Dr Glanville, Dr Reginald Stradling (later Sir Reginald) Chief Scientific Advisor to the Ministry of Home Security, Mr Horace Morgan, Professor Desmond Bernal, Professor W.R. Thomas and Barnes Wallis himself.

Towards the end of 1941 Dr Stradling, with the consent of the Birmingham Corporation, suggested to the Committee that the small Nant-Y-Gro dam, built at the end of the last century to provide water for the construction of the Elan Valley dams, and no longer required, might be used to test the gravitational effects of scale. It was about 9 metres high and 55 metres wide, of the straight gravity design and made of mass concrete. Built in a remote

---

[1] This Committee played a helpful role in subsequent events and must not be confused with the fictitious and very hostile committee portrayed in the 'Dambuster' film.

locality, it could be breached safely, with the water certain to run harmlessly into the Caban-coch reservoir.

Nant-Y-Gro was in an almost inaccessible spot and the bomb had to be manhandled to the location. With the help of a few Royal Engineers using a small boat, the bomb was finally suspended three metres below the water level. Two high-speed cinematograph cameras and their operators were provided by the Royal Aircraft Establishment, and arrangements were made to record pressures and movements. The test was successful and gave confidence for further tests.

Meanwhile, Wallis, after a close study of the Lancaster drawings, had decided that a large spherical bomb could be carried. It was now vital for him to enlist the interest and help of some eminent people. One such person was Sir Henry Tizard AFC, a former Royal Flying Corps and Royal Air Force test pilot who became a Colonel in the RAF Experimental and Design Unit, and was now with MAP. Sir Henry proved very helpful to Wallis, acquiring the use of the water tanks at the National Physical Laboratory at Teddington, where Wallis spent many daylight hours firing wooden balls of different weights and sizes from a catapult, out along the surface of the water. Until he could accurately calculate ratios of weight, size and speed to length of bounce, any further progress would be limited.

Tizard came to watch the experiments, and after two days a pattern of bounce began to emerge. Wallis finally managed to put a back spin on the ball and came near enough in his calculations to predict, with some accuracy, the outcome of a bouncing bomb attack on a dam, from the air.

Finally the time came for testing with the real thing, but to do this he had to have Government backing. Fortunately, Group Captain Winterbotham knew G.M. Garro-Jones (later Lord Trefgarne), another former RFC pilot, who was Parliamentary Private Secretary to the Minister of Production. A new committee had recently been set up under the Ministry to examine new projects of this kind. Its Chairman was Sir Thomas Merton, formerly Professor of Spectroscopy, at Oxford University, and now Scientific Advisor to MAP. Garro-Jones was so impressed by the notes of Wallis' invention that he obtained a hearing with Merton's

Committee, resulting in the go-ahead to construct and test a prototype bouncing bomb.

Prototype bombs were constructed and initial tests were made with a converted Wellington bomber. The first of these took place on 4th December 1942, over Chesil Beach, near Weymouth, Dorset. The aircraft was piloted by Mutt Summers, Chief Test Pilot for Vickers, with Wallis acting as bomb-aimer. They flew low over the water and at the appropriate moment, Wallis pressed the release button. The bomb hit the water but its case crushed slightly on impact. It would have to be made stronger.

Eight days later they tried again, and this time it worked. Over the next few days they dropped three more bombs, the tests being recorded by movie camera.

Early in January 1943, the Managing Director of Vickers Armstrong, Sir Charles Craven (a former Royal Navy Commander) told Wallis that he was to go to London as the First Sea Lord, Sir Dudley Pound, wanted to see the films of the Chesil Beach drops. He was seated with the Admiral, and the film started, with the title *Most Secret Trial Number One*. Sir Dudley watched the Wellington bomber with its bulge beneath it, saw the bomb drop away with a backward spin at considerable speed and fall towards the sea. When it hit the water it bounced about a dozen times and covered about a half mile over the surface before it finally sank.

The Admiral was impressed, but it transpired that the Navy were eager to use such a weapon against the German battleships. Wallis quickly realised that if so used, all hope of surprise in its planned use against the dams would be lost. His only course was to interest the Prime Minister sufficiently for him to order the dams raid a priority. To this end he again contacted Professor Lindemann, who by now was Lord Cherwell.

Wallis wrote to him on the last day of January with a twenty page 'secret' report, complete with photographs, diagrams and the theory of the spinning bomb. He suggested that the best time to attack the dams would be in mid-May 1943 if the moon was full and not covered by cloud. There followed a meeting with Cherwell and the film was shown again, but there was no result from this

Copy of an original drawing showing
the principle of the bouncing bomb.

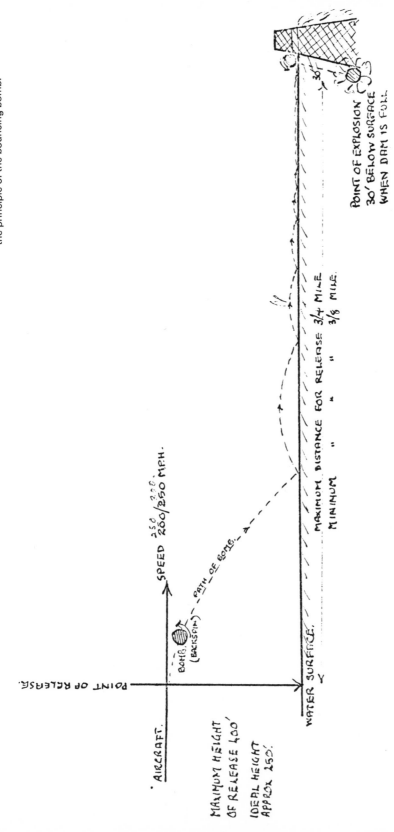

meeting, and no indication that any further action was forthcoming.

At a meeting at Air Ministry on 13th February, note was made that a spherical bomb – now given the code-name of 'Upkeep', was being developed and that trials were taking place with full mock-up models. These trials indicated, the Committee heard, that the weapon could be successful if used. The size of the bomb used in the trials was approximately 36 inches in diameter and its weight around 1,000 pounds. Tests about to be made would be carried out with the use of a specially modified Mosquito aircraft.

A further meeting was arranged for the 15th, so that further evidence could be obtained from further tests at the Teddington tanks. This time a two-inch ball would be fired at a model of the dam while it was filmed by a young lady housed in a water-tight tank below the water level. This test, the results of which were presented to the Committee, showed the ball finally plunging below the surface and crawling into position against the side of the tank.

Meanwhile, on the 14th, an internal letter was sent to the Commander-in-Chief of Bomber Command, Sir Arthur Harris, from his SASO, Air Vice-Marshal Robert Saundby, concerning the meeting on the 13th. He outlined the plan to attack the German dams with a bouncing bomb, and concluded that he thought it was entirely possible. He also suggested that one squadron of bombers should have its aircraft suitably modified to carry the weapon and used to attack the dams.

Harris replied the same day, or rather handwrote on the letter which he returned to Saundby:

This is tripe of the wildest description. There are so many ifs and ands that there is not the smallest chance of it working. To begin with the bomb would have to be *perfectly* balanced round its axis, otherwise vibration at 500 RPM would wreck the aircraft or tear the bomb loose. I don't believe a word of its supposed ballistics on the surface.

It would be much easier to design a bomb to run on the surface, built to nose in on contact, sink and explode. This bomb would of course be heavier than water and exactly fit existing bomb bays.

At all costs stop them putting aside Lancasters and reducing our bombing effort on this wild goose chase. Let them prove the practicability of the weapon first. The war will be over before it works – and it never will.

ATH 14/2

To be fair to Sir Arthur, it should be remembered that he had many ideas put to him while C-in-C Bomber Command, many of them far-fetched. Everyone seemed to be an expert in bombing Germany. One scheme worthy of mention concerned an idea of bombing the *Tirpitz* while anchored in Aasfjord, near Trondheim, Norway. The attack was by a force of Stirling and Halifax bombers, led by Wing Commander Donald Bennett, later Air Vice-Marshal and leader of 8 Group, Bomber Command. The object was to drop ordinary sea-mines, minus their horns and fitted with hydrostatic fuses, set to explode at thirty feet below the surface of the sea. They were dropped at low altitude to roll down the side of the fjord and sink under the battleship. Bennett's Halifax was set on fire by the flak defences and he and his crew had to bale out. He succeeded in escaping to Sweden with his radio operator and returned to England.

A reconnaissance Mosquito found no apparent damage to the *Tirpitz* and it was later confirmed by members of the crew that only some paint was scrapped from the ship's bottom, but enough fish were killed to keep them fed for several days!

Harris was also concerned that he should not lose any of his precious Lancaster bombers in any hair-brained schemes and in this he was supported at this time by the Chief of Air Staff, Sir Charles Portal. Portal had also seen the Chesil Beach films and wrote to Harris assuring him that he would not allow more than three Lancasters to be diverted on tests for the Upkeep bomb.

Then, on 22nd February, Harris was advised that Wallis would be visiting him with the film. He arrived with Mutt Summers, and set out to explain his ideas after which Harris seemed to be a little more interested and agreed to watch the film. For reasons of security only Harris, Wallis, Summers and Saundby were present in the Command's projection room, Saundby working the projector. At the end of the film all Harris agreed to do was to think it over.

The very next day Wallis was sent for by Sir Charles Craven[1] at Vickers House, and told to stop all ideas of attacking the dams. On hearing this shattering news, Wallis proffered his resignation and left. He went straight to Sir Thomas Merton, told him he had resigned and that the dams raid was off. Merton's assistant, Sydney Barratt, then questioned Wallis for ninety minutes on the whole scientific basis of his case. The outcome was that the Prime Minister was sent the Upkeep papers. After reading through them, Churchill, gave the order for a raid on the German dams to be prepared.

The news came as a great relief to Barnes Wallis after months of trial. His resignation to Craven had only been given verbally and was not accepted and he was soon back in his Weybridge office in order to begin the next round of tests. This bomb had still to be perfected and it was now nearly March – less than two months away from his projected raid date of mid-May!

With all obstacles brushed aside by the Prime Minister's order, a plan of attack had now to be drawn up on the assumption that a bomb would be ready in time. Group Captain C.E.H. Verity, in charge of Airforce Intelligence (AI3(C)), which prepared the information for all fighter and bomber operations in Europe, received a call on his green scrambler telephone. It came from his boss, Air Commodore Patrick 'Tubby' Grant, Deputy Director of the Intelligence Office, who informed him to expect a civilian gentleman named Wallis to call, in order to discuss dates for a very special operation. He was to be given every assistance and information he required, including very secret material.

Shortly afterwards, a dishevelled Wallis, with his distinctive shock of white hair, bounded into Verity's office with great enthusiasm, escorted by a guard, and immediately explained the outline of the operation and his bomb. Verity's desk was soon covered with sheets of sketches etc, which he found not only

[1] Managing Director of Vickers and attached to the Ministry of Aircraft Production.

absorbing but quite brilliant. As Verity had been a Chartered Civil Engineer before the war, Wallis assumed he would understand the massive engineering work behind the construction of the dams. Luckily, Verity had three very highly qualified people on his staff, and was able to settle them in the Library of the Institute of Civil Engineers where a vast number of volumes dealing with dams and German civil engineering works were kept.

They quickly tracked down the papers and articles on the Möhne dam, which revealed minute details of the design and construction. They began a lengthy correspondence with Barnes Wallis who seemed to appreciate greatly the work being done for him and the material he was being supplied with. All letters were sent via Effingham golf club, and no contact was made with him at either Vickers or his private address. Group Captain Verity then got down to producing some details of the standard target material, maps, information sheets, drawings, photographs, etc.

On the afternoon of 26th February, Wallis, Verity and Sir Charles Craven, were summoned to the office of Air Marshal F.J. Linnell OBE, whom Wallis had met during the early days of his experiments. He told them that the War Cabinet had directed that the testing and development of the bomb with a modified Avro Lancaster to carry it, was to proceed at once. 'The Air Staff have ordered,' he said, 'that you are to be given everything you want.'

Linnell promised three Lancasters modified by Avros.

Barnes Wallis was now working between his drawing office at Surhill, Woolwich, where the bombs were being filled, and the experimental dropping ground at Reculver Bay, near Margate in Kent. It had been a long hard road for Wallis to reach even this stage, having to overcome all manner of obstructions. But he had persevered, determined to convince all the sceptics that his idea was not only possible but now an almost certainty.

CHAPTER TWO

The Target ...
The Weapon ...

The plan was to attack and breach all, or as many as possible, of the five main German dams. By far the most important was the Möhne dam and hydro-electric power station at Gunne, six and a half miles south-west of Soest, and twenty-five miles east of Dortmund.

It is situated at the north-west corner of a reservoir which extends eastwards from the dam for about six miles and south and east for about three and a half miles. The dam had been built to hold a storage capacity of between 13 million and 140 million cubic metres of water and was constructed between 1909 and 1913, in order to improve the flow of the River Ruhr, by helping with water shortages during summer and autumn. In addition, the numerous pumping stations which provided water for the towns of the Rhine-Westphalia area, as well as hydro-electric plants along the river, could be supplied with adequate water even during dry periods.

The German armament industry was very dependent on the water (to produce one ton of steel required 10,000 tons of water). This may also be gauged from the fact that coal mines require about one cubic metre of water per ton of coal raised; coke ovens about two cubic metres, while blast furnaces need two cubic metres for each ton of iron produced. Power stations and chemical industries also make heavy demands on water supplies. The dam, was therefore, of vital importance to the whole Ruhr Valley and its destruction would not only cause disastrous flooding but would upset water supplies over a large part of the most mighty industrial area of Germany.

The dam wall was made of limestone rubble masonry and was specially protected against seepage of water. It was bedded onto at least six and a half feet of rock at its base and was built in a curve which permitted a good joining of the ends of the wall to the sides of the valley.

The Möhne is 130 feet high from its rock bottom. The maximum height of the dammed water being 105 feet. Its length is 2,100 feet and its thickness is 25 feet at the top, 112 feet at the bottom. The water was normally drawn off by means of four pipes, 4.62 metres in diameter, carried in pairs through culverts in the foot of the dam. Each pipe had two valves, controlled from the crest of the dam and operated by either electric power or by hand. The control rooms were in the round chambers built on each side of the gabbled towers on the top of the dam, one branch of each serving to discharge the water directly into the Möhne and the other carrying it to the power station. This station was directly below the dam, operated by the Vereinigte Elektrizitaetswerke Westfalen AG, (VEW) equipped with two driving turbines, a main turbine of 2,200 hp, and an auxiliary turbine of about 1,000 hp, at maximum output.

Gravity dams such as the Möhne offer complex problems as a target owing to the massive masonry, or concrete, walls. These are approximately triangular in cross section, resisting the great press of water with its tendency to try and push them downstream, and also cause a problem through their sheer weight and the fact that they are keyed to the rocky bed of the valley across the outlet from which they stand. The immense strength of the dam can be imagined from its thickness of 112 feet at its base. When the reservoir was full, the depth of the lake in the middle was about 128 feet.

Along the River Ruhr there were approximately twelve dams with a total capacity of 200 to 266 million cubic metres. In 1943 the Möhne dam was estimated to contain 134 million cubic metres, or approximately half of the total of the twelve dams.

The dam did have some defences. Two rows of torpedo netting hung down from floating wooden beams and stretched across the water, 100 to 300 feet out from the top of the dam wall. In August 1939, the German Mayor, Duigardt, made no secret of his concern over the lack of military defences for the large Westphalian dams. In a letter to the military authorities in Munster, he pointed to possible results should the dams be destroyed from the air, not so much from a direct hit, but by a bomb dropped 20 or 30 metres from the dam, which would explode below the water line! Under pressure of many such letters as this, the military relented and at the outbreak of war flak defences were installed at the Möhne. When the offensive in the west began in May 1940, the guns were removed to protect, so the authorities believed, a more worthwhile target for British bombers, but in the autumn of 1942 they once again appeared. Barrage balloons were also put above the dam but were removed and not replaced.

The limestone used in the construction was a kind found in the Ruhr Valley near Neheim Hüsten, and also a kind of sandstone called graywacke, with a small proportion of green sandstone. The mortar was a cement-trass in proportion of one cement, three white lime, five trass and twelve sand. Unlike the construction of other dams, two rock-sand and not river sand was used, ie: crushed stone. Great importance was placed on impermeability. On the water side an impermeable coating of half limestone, two and a half cement, one and a half trass and six Rhinestone with 2% asphalt emulsion coating, was applied. The final smooth coating consisted of one cement to two sand with emulsion added. The coated surface was then painted over with two coats of 'Siderosthen' or 'Nigrit'. To protect it against waves, frost, heat of the sun and mechanical damage, the plastered surface had a facing of sixty to ninety centimetres to the full height. Further, a wall of clay, two metres thick was rammed half-way up the wall from the valley bottom.

The second target in the Ruhr Valley area, was the Sorpe dam. It was built between 1927-1933, of earth with a concrete core. It had a water volume of 72 million cubic metres. The height of the wall is 58 feet. It is situated on the River Sorpe, a tributary of the Ruhr and six miles south of the mighty Möhne dam. It took three years to fill with water after its completion. The dam is closed by an earth and stone wall with a centre cement core at the base of which there is an inspection gallery. Although chosen as a target, it was not, in the opinion of Barnes Wallis, suitable for his bouncing bomb, being of solid concrete.

The third dam selected to be attacked was the Eder. Like the

Sorpe, the Eder is situated to the west of the Ruhr. This dam is situated in the area of the River Eder and built of masonry. Its water volume is 202 million cubic metres and was built at Hemforth, about two miles south of Waldeck, twenty-four miles north-west of Bad Wildungen. Its power station was the 30,000 KW Hemforth I and II hydro-electric stations of the Preussische Elektrizitaets AG.

The Eder is the largest dam in Germany, being 139 feet high, 1,310 feet long with a thickness of 19 feet at the top, 115 feet at the bottom. Its reservoir area was much larger than the Möhne and water capacity was greater in consequence – 7,100 million cubic metres. Its prime purpose is to store water for protection against flooding and also for increasing the water flow of the Fulda and Weser Rivers, and the Mittelland Canal, in order to improve navigation. It was believed that the destruction of the Eder would probably result in the destruction of the four associated hydro-electric Power Stations at Hemforth, (2), Brenghausen, and Offolden, which together had a total capacity of about 150,000 KW's which constituted an important item in the electrical supply system of the Preussen Elektra, particularly the Brenghausen Station which was a pump storage peak load plant. The dams' destruction would result in serious flooding and damage to communications and industrial plants on the banks of the river. The Sorpe and Eder were used in conjunction with each other.

The fourth dam was the Ennepe, built between 1902-1905. It is sometimes referred to as the Schelme dam as it is built on the Schelme, in the Arnsberg area. Its water volume is 15 million cubic metres, and its walls are of masonry. The height of the wall is 45 feet. The fifth dam, was the Lister, built between 1909-1911, again built of masonry to a height of 35 feet. Built at Attendorn, its capacity is 22 million cubic metres of water.

Wallis' bomb or mine (both words were used to describe it) was sixty inches long and fifty inches in diameter. It had three hydrostatic pistols set to detonate at 30 feet, also a ninety-second time fuse, initiated at release, intended to destroy the weapon if for any reason the pistols failed to function. On each end of the cylindrical casing was a hollow circular tract, twenty inches in diameter into which disc-like wheels mounted on supporting calliper arms were fitted with fore and aft axis.

The total weight of the bomb was 9,250 pounds, consisting of 6,600 pounds RDX Charge Weight and 2,650 pounds of bomb casing, mechanism, etc. A safety pin was attached to a wire which in turn was attached to one of the calliper arms, to release the bomb when the bomb-aimer pressed his release button in the normal way. When activated, the calliper arms holding each side of the bomb, sprung open, forced open by powerful springs, and away went the bomb, having already been set in a backwards motion several minutes before release. In the same moment the safety pin was automatically removed.

The best way to describe the Wallis bomb would be to say it looked like the front wheel of a steam roller. The external diameter of the discs were slightly smaller than the weapon itself and the power for rotating the weapon before release was provided by a VSG hydraulic motor with belt drive, to one of the discs. The desired spin was then transmitted to the bomb by friction contact between the internal tracks and the driven disc. The motor was started ten minutes before the arrival at the target and rotation controlled at 500 RPM. The calliper arms were held inwards by a straining system, retained by a bombslip.

The height of release had to be 60 feet, at a speed of 240-250 MPH, and it had to be released at between 400 and 500 yards from the dam wall.

CHAPTER THREE

No 617 Squadron is Formed

While Wallis was left to conclude his trials, Sir Arthur Harris was left to plan the next stage in the operation. On 15th March he sent for Air Vice-Marshal Ralph Cochrane, who was Air Officer Commanding (AOC) No 5 Group of Bomber Command, whose Headquarters were at St Vincent's Hall, Grantham, Lincolnshire. Harris told Cochrane of the Wallis bomb and the proposed attack, and that he would like a squadron from his Group to undertake the task.

Cochrane felt that such an attack as outlined would take experienced bomber crews and that it would be far better to form a new squadron from men from the Group. In the main these would have to be men or crews who had just finished or were about to finish a tour of operations, a tour at this stage in the war being anything from twenty-five to thirty missions. Harris then asked Cochrane who he thought should command such a squadron and lead the attack. To this Cochrane replied, 'Wing Commander Guy Gibson'.

This was on 11th March, and that night Gibson flew a raid to Stuttgart as his last operation of his third tour – two bomber and one on night fighters. At twenty-four years of age, Guy Penrose Gibson had a wealth of experience behind him. Born in India on 12th August 1918, where his father was a Conservator of Forests, he was educated at St George's Prep School, Folkestone and St Edward's School, Oxford, his family, parents and one brother and one sister, living in Porthleven, Cornwall. Gibson joined the RAF in 1936, being granted a short service commission. Following his pilot training he was posted to No 83 Squadron in September 1937. He was with 83 when war was declared. By 1943 he was in command of 106 Squadron.

His final operation on 15th March was far from straightforward. En route to Stuttgart his aircraft developed engine trouble which caused a gradual loss of height. He decided to try and complete the mission if possible, which he and his crew did, returning safely to their base at Syerston, Nottinghamshire, but only after hedge-hopping back over Germany and France. On his return his immediate thoughts were of a holiday in Cornwall with his wife Eve. With them would go Gibson's black labrador dog, 'Nigger'. Nigger had been with him in 106 Squadron and had often flown with his master. He was a favourite with almost everyone and in the Mess could be seen slurping away at a pint of beer that some kindly soul would place on the floor for him. The holiday, however, was not to be. Gibson was posted to Group HQ at Grantham and a few days later the AOC Ralph Cochrane, sent for him.

'How would you like to do one more trip?' asked the AOC.

'What kind of trip, Sir?' asked Gibson. Cochrane could not be specific at this stage – indeed it was some time before Gibson was finally told the target – but Cochrane stressed the importance of the task, so Gibson agreed to take on the job. All he really knew was that a new squadron would be formed, he was to command it, lead it, and have an almost free hand to pick his pilots – something very few commanders ever had the chance to do.

On 17th March 5 Group HQ received a letter from Bomber Command, telling them of a special weapon, code-named Upkeep, which had been produced. It was intended for a special mission against a large dam in Germany. The attack would have to be made during the full moon period of May, as after this time the water level in the dam would be too low for the weapon to be effective.

At about this time, Cochrane sent for Wing Commander W.E. 'Wally' Dunn, a regular air force officer who had served with Arthur Harris on airships at Howden, 1920-21. Wally, a Yorkshireman, was Chief Signals Officer at 5 Group. Cochrane explained to him the proposed operation and that he required the new squadron to be provided with everything they needed in the way of communications and radar. Dunn was also made aware that the operation was top secret and had, what was termed as a

'Churchill Priority'. The cover plan was that it was an attack on the *Tirpitz*.

Cochrane and Gibson met again on the 18th. Also present was Group Captain Charles Whitworth, Station Commander at RAF Scampton, Lincolnshire. Scampton, just a few miles north of Lincoln, had been chosen as the new home for Squadron 'X', as it was still known. This was changed to 617 Squadron on the 24th. The only other squadron at Scampton was 57 Squadron, also flying Lancaster bombers, and they provided some of the men for the new squadron.

One man who came was Sergeant Jim Heveron, as Orderly Room Sergeant. After his first meeting with Gibson, Heveron said that he seemed a well-balanced leader with the inimitable quality of knowing his job. Heveron later helped Gibson with his now famous book *Enemy Coast Ahead*, which Gibson wrote while at Air Ministry, sending for his former Orderly Room Sergeant to assist him.

Heveron had been at Group HQ when posted back to Scampton, arriving early one morning and reporting to the old 49 Squadron hangar. He could find nobody in the offices, but found a queue of ground staff of all shapes and sizes, seemingly waiting for someone to tell them what to do. He finally located Flight Sergeant G.E. Powell, also posted in from 57 Squadron; then he started to scrounge paper, furniture and various other odds and ends needed to set up an orderly room. The queues of men lasted all day and by the evening some 250 airmen had been booked in and accommodated, which under normal conditions would have taken two months.

'Chiefy' Powell, as Administration NCO, was to prove a tower of strength to Gibson and 617. From Wrexham, he had been named George after King George V, and Edward after King Edward. He had joined the RAF in 1928 and had served as an air gunner in the early part of the war, flying with Coastal Command. He admired Gibson very much, saying that he was the finest officer he had ever served under. He was to remain with 617 for the rest of the war, and served with the RAF until 1955.

Jim Heveron got the Orderly Room sorted out, despite some poor staff sent in by other squadrons, but had some valuable help from a WAAF who was loaned to the squadron for two days.

Heveron found Gibson a pretty tough CO where anything connected with flying was concerned. He remembers Gibson's office was painted out in sky blue with seagulls and aeroplanes all over it, and he always kept a service revolver on his desk.

On taking over as CO of 617 Squadron, Gibson's first task was to select twenty-one pilots and crews – 147 men in all. The rapid formation was only possible by the existence of a group pool of ground and handling equipment that was already at Scampton.

The nucleus of ground crew came from 57 Squadron. As well as Chiefy Powell, there came Corporal John 'Jock' Bryden, a Scot from Ayr. He was an engine fitter and worked in the engine and power plant section. From Gibson's old squadron, 106, came airman mechanic Payne and engine mechanic Nick Carr, who whilst on 617, serviced both Gibson's and Squadron Leader David Maltby's Lancasters. Some ground crews were posted to 617 on the invitation of the new aircrews themselves, itself something new in the RAF. One of them was Leading Aircraftman Lofty Stretch, a flight mechanic engineer from 50 Squadron, invited in by Flying Officer Les Knight, one of Gibson's selected pilots. The ground crew men were always the backbone of any squadron and any measure of success that 617 Squadron obtained, was due in part to their devotion and hard work.

The aircrews selected by Gibson began to arrive on 24th March. Twenty other crews were required, the men arriving during the period 24th March to 17th April. Of the pilots, four came from 57 Squadron, three from 97, four from 50, three from Gibson's 106 Squadron, two from 49 and one each from 44, 61, 207 and 467 Squadrons. With Gibson's own crew this totalled twenty-one operational Avro Lancaster crews.

Ralph Cochrane had had little hesitation in recommending Gibson for the command of 617 Squadron. He was a very experienced pilot, bomber captain and leader. His war had begun with 83 Squadron in 1939; in fact he flew his first operational sortie on the early evening of 3rd September 1939, but it proved a disappointment, the pilots having to jettison their bombs into the sea when they failed to find German naval targets.

Guy Gibson and Nigger.

Officer Burpee and crew.

Officer Knight and crew.

With the 'phoney war' period, it was not until 11th April 1940 that he flew his next mission, but by September of that same year he had brought his mission total to twenty-seven, having earlier received the DFC. Leaving the Hampden-equipped 83 Squadron, he served as an instructor until November when he was posted to No 29 (Night Fighter) Squadron for a 'rest'. Flying in the night skies over England during the Blitz, he shot down three German raiders, probably destroyed one and damaged four more. He completed ninety-nine night sorties, gaining a bar to his DFC.

Following a further spell as an instructor, this time as Chief Flying Instructor at No 51 Operational Training Unit (OTU), he was given command of 106 Squadron at Coningsby in March 1942, flying firstly Avro Manchesters, then the new Lancasters. He led 106 valiantly for a year, during which time he won the DSO and brought his number of bombing operations to seventy-three, having flown no less than forty-six ops with 106 – almost the equivalent of two tours! There were very few better qualified to command 617 in the spring of 1943.

As he completed his tour with 106 Squadron, Ralph Cochrane recommended:

Wing Commander Gibson has now handed over command of 106 Squadron after an outstandingly successful tour as squadron commander. In view, however, of the recent award of the Distinguished Service Order, I recommend that he should now be considered for the award of a second bar to his Distinguished Flying Cross, rather than a bar to his DSO.

To this, Sir Arthur Harris added: 'Any Captain who completes 172 sorties [73 bomber and 99 night fighter] in outstanding manner is worth two DSOs if not a Victoria Cross.' A bar to his DSO was duly awarded.

The three pilots posted in from 97 Squadron were, Flight Lieutenant J.C. McCarthy, Flight Lieutenant J.L. Munro and Squadron Leader D.J. Maltby, all of whom arrived on 24th March.

Joe McCarthy was a 23-year-old American from New York, born of Irish-Swedish parents. At over six feet tall and 15 stones in weight, little wonder he was known as 'Big Joe'. He had joined the RCAF in 1941 and whilst with No 14 OTU in 1942 had taken part in the Thousand Bomber raids on Cologne and Hamburg. With 97 Squadron at Woodhall Spa, he had just completed a 30-operation tour and had been recommended for the award of the DFC. During his tour he had flown to Berlin three times and to Italian targets five times.

Les Munro was from New Zealand where he had been a farmer. He was approaching his twenty-fourth birthday and during his tour with 97 Squadron he, like McCarthy, had been to Berlin three times. He, too, had recently been recommended for the DFC.

David Maltby was twenty-three and from Sussex, England. He had joined the RAF in 1940 and a year later went to 97 Squadron. He received the DFC following completion of twenty-seven operations in July 1942. He had just returned to the squadron to begin his second tour when he was called to 617.

Two other pilots who arrived on the 24th were the first from 57 Squadron at Scampton, Flight Lieutenant William Astell and Flying Officer G. Rice from 50 Squadron. Bill Astell hailed from Manchester, joining the RAFVR soon after the outbreak of war. Following training in Rhodesia and the Middle East, he flew Wellingtons from Malta, winning the DFC, and later in North Africa. Returning to England in 1942, he joined 57 Squadron in January 1943. Geoff Rice was twenty-six, having joined the RAF early in 1941. After service with 50 Squadron he too was ordered to join the new 617 Squadron.

Flight Lieutenant J.V. Hopgood was a Londoner, aged twenty-one, although his early life was spent in Surrey. He was the first of the contingent to come from 106 Squadron. He had received his pilot training at RAF Cranwell, winning his wings in 1941. Initially he flew with 50 Squadron, although his first ten ops were flown as navigator. Then in April he joined 106 Squadron as pilot and captain of a bomber crew. With 106 he won the DFC following thirty-two operations, and Guy Gibson himself recommended a bar to this decoration at the end of 1942 following the completion of forty-five operational sorties. A gentle man, his reason for joining Bomber Command was so that he would not be able to witness human suffering from man's weapons of war, although he knew he had to fight for his country. In his younger days he had mistakenly shot a woodpecker instead of a pigeon. He had been so distressed that he had it stuffed and enshrined in a glass case in his room.

Another 106 Squadron pilot to join 617 at the end of March was Flight Lieutenant D.J. Shannon DFC. Dave Shannon was twenty, but looked younger. He came from South Australia where he had been a bank clerk. After his training he was posted to Gibson's squadron in 1942 and in July actually flew three operations as second pilot to Gibson. He was awarded the DFC at the end of that year having flown twenty-six ops – many being against tough targets over both Germany and Italy. He joined 617 with a total of thirty-six missions.

Two other Australians came in from 50 Squadron, Flying Officer L.G. Knight on the 26th, and Flight Lieutenant H.B. Martin DFC on the 31st. Les Knight, from Victoria, was just twenty-two years old and had enlisted in the RAAF in February 1941, joining 50 Squadron as a sergeant pilot in September 1942 but received a commission in December. By the time he joined 617 he had been recommended for the DFC having completed twenty-six missions, but the events of the Dams Raid earned for him a higher decoration.

Micky Martin was twenty-five, from Edgecliffe, New South Wales, but had lived in Sydney. He initially flew Hampdens with 455 RAAF Squadron in late 1941 before going to 50 Squadron in April 1942. He ended his period with these two units with a total of thirty-six ops, thirteen with 455, twenty-three with 50 Squadron.

The two pilots posted in from 49 Squadron were Flight Sergeant W.C. Townsend DFM on the 26th, and Flight Sergeant C.T. Anderson on 17th April. Bill Townsend was twenty-two, from Gloucester and had been in the army before transferring to the RAF. Receiving his wings in December 1941, he joined 49 Squadron in June the following year, although he, like McCarthy had been roped in to fly on the Thousand Bomber raids (to make up the numbers) while still at OTU. He left 49 with twenty-six missions flown and the DFM. Among his raids had been a daylight trip to Milan on 14th February 1943, while earlier, on 23rd January, he brought home his Lancaster on three engines. Cyril Anderson came from Wakefield, Yorkshire, and had married two

(*Left*) Bill Townsend CGM, DFM. (*Right*) John Hopgood DFC.

(*Left*) Vernon Byers. (*Right*) Squadron Leader Henry Young DFC.

weeks after the war began. His early years in the RAF were in engineering, but he trained as a pilot in England and Canada when war started. He joined 49 in February 1943, but almost immediately joined 617. He was, however, commissioned later, but had only flown eight missions.

Three Canadians joined the squadron at the end of March: Flight Sergeant K.W. Brown in from 44 Squadron, Sergeant V.W. Byers from 467, and Flying Officer L.J. Burpee, the latter also in from Gibson's old 106.

Ken Brown was from Moose Jaw, Saskatchewan and had flown just seven ops with 44 Squadron between 5th February 1943 and his posting to 617. He was twenty-three years old. Vernon Byers was also from Saskatchewan, and a good deal older – he was thirty-two, joining the RCAF when he was thirty. He went to 467 Squadron in February 1943 and had completed four or five ops before the move. He once wrote to his mother, Ruby, that when the war was over he would bring her to England and fly her over the countryside so that she too could see how beautiful it was.

Lewis Burpee, just twenty-five when he joined 617, came from Ottawa. In May 1941 he graduated from Queen's University, Kingston, Ontario, with a BA Degree in English Literature. He loved music and reading, indeed, his mother was a member of the Ottawa Symphony Orchestra. Enlisting in the RCAF in December 1940 while still at University, he received his wings and came to England in September 1941. He had wanted to be a fighter pilot but to his surprise found himself flying twin-engined Whitleys. In England he met his wife to be, Lillian, who worked in Oxford, having been evacuated from London. They were married in September 1942. Soon afterwards he was posted to 106 Squadron, his wife obtaining two rooms near to the airfield at Newark. Towards the end of his tour, after flying twenty-six ops, he received the DFM. He had had an interesting tour, having flown to Milan in daylight, and twice having brought back shot-up aircraft. He was commissioned, and soon after posting to 617 found that his wife was expecting their first child.

Flight Lieutenant Robert Barlow DFC arrived on the 31st, posted in from 61 Squadron on being tour-expired. Twice he had brought back aircraft on three engines, from Stuttgart in November 1942,

(*Left*) Stan Hewitt, Danny Walker, Bob Hutchinson. (*Right*) Sidney Hobday DFC .

(*Left to Right*) Sergeants 'Jock' Peterson, Cyril Anderson, Doug Bickle, Arthur Buck, Jimmy Green, John Nugent.

and Berlin on 1st March 1943. He had also brought back damaged aircraft from Turin and another Berlin trip. He was approaching his thirty-second birthday.

The flight commander for B Flight arrived on 31st March. This was Squadron Leader H.E. Maudsley DFC from Nottingham. He joined the RAFVR from Eton College in 1940 and in May 1941 joined 44 Squadron, flying Hampdens, completing a successful tour at the end of September when he received the DFC. After more than a year on a Conversion Unit, he joined 50 Squadron for a second tour in March 1943 but was soon posted to 617 Squadron. Air Vice-Marshal John Slessor, when he had been AOC of 5 Group, wrote that Maudsley was among the outstanding Captains in his group.

On 6th April Pilot Officer W. Ottley arrived from 207 Squadron. Warner Ottley had enlisted in the RAF in 1941 and following a brief spell with 83 Squadron in mid-1942 had flown with 207 from that November. With them he completed a tour of 31 operations, then joined 617 for one more mission. The final three pilots in from 57 Squadron joined in the second week of April They were Flight Sergeant W.G. Divall, Squadron Leader H.M. Young and Flight Lieutenant H.S. Wilson.

William Divall came from Surrey and was twenty-two. He trained in Canada and went to 57 in February 1943 but was not with them long before joining 617. Henry Young was another American and came to 617 as A Flight commander. He was twenty-six a Californian and had been educated in England at Westminster and then at Trinity College, Oxford, where he read law. He also rowed for the Oxford team in the 1938 Varsity Boat Race, as No 2 in the crew. Oxford won that year by a distance of two lengths, in 20$\frac{1}{2}$ minutes. He was also a member of the University Air Squadron. In June 1940 he joined 102 Squadron at Driffield in Yorkshire, completed a tour of operations and won the DFC. On two occasions, firstly in October, he was forced to ditch in the sea, a dip that lasted twenty-two hours before rescue. It was undoubtedly his insistence on dinghy drill that saved him and his crew (not to mention his Oxford rowing training!). A second sojourn in a dinghy came in November after a trip to Turin, and earned for him the nickname of 'Dinghy' Young. Following a spell as an instructor he joined 104 Squadron, completing a second tour, including a period in the Middle East when in June 1942, the squadron was sent to Malta. Here Young continued to fly sorties against Axis targets in Italy and North Africa. Shortly afterwards he received a bar to his DFC, having flown over fifty operations.

A tale is related of him by Wally Dunn that 'One day his flight pilots came to see me as Station Signal Officer and asked if I would teleprint a signal from HQ Bomber Command to RAF Driffield posting Flight Lieutenant H.M. Young to command a new dinghy training school at Calshot. Following a custom of mine to do anything for my aircrew I duly typed it out and stamped it with various signal marks, and put it with the signals for the day. It worked very well to the extent that 'Dinghy' bought everybody a drink in the mess before the truth was known.'

The pilots did not necessarily arrive with their crews. Some came with part, perhaps one or two, others with none. Gibson finally had his 147 men, and began to sort out who wanted what in order to end up with twenty-one seven-man crews who would fly and train together on the proposed attack.

As can be seen from the list of pilots, they had a diversity of experience. Several were vastly experienced, others still learning their trade as bomber captains. All, however, apparently passed Gibson's standards, standards he felt were necessary for a precision raid on a group of unusual targets at low level. As Gibson recorded in his book, *Enemy Coast Ahead* (Michael Joseph Ltd) he took an hour to pick his pilots, writing all the names down on a piece of paper and handing it to the appropriate person at 5 Group HQ, who was charged to make the necessary postings. Those he selected he chose because he had personal knowledge of most of them and believed them to be the best bomber pilots available. The crew members were more difficult but help came from the aircrew officers, and soon Gibson had his squadron.

Guy Gibson's own crew arrived in ones and twos. Navigator was Flight Lieutenant T.H. Taerum from Canada, who had just completed a tour with 50 Squadron, coming in from a Conversion Unit at Wigsley. Flight engineer was Sergeant John Pulford from Hull, who had flown ten missions with 97 Squadron. The wireless operator was Flight Lieutenant R.E.G. Hutchinson DFC, from Liverpool. George had won his DFC with 106 Squadron, and was the only one of Gibson's own crew to follow him to 617 after a spell with a Conversion Unit on 'rest'. 'Hutch' is remembered by another of Gibson's 106 crewmen, Brian Oliver, as one who never saw flak coming up as he always kept the curtains of his wireless cabin firmly closed. He was also prone to severe air sickness but in no way did he let this deter him from flying trips over Germany.

Flying Officer F.M. 'Spam' Spafford DFM, bomb aimer, was from Adelaide, Western Australia. As an NCO he flew with 50 Squadron, receiving an 'immediate' award of the DFM in the autumn of 1942 following fifteen raids. He ended his full tour in January 1943. Rear Gunner was Flight Lieutenant R.A.D. Trevor-Roper DFM, from the Isle of Wight. Before joining the RAF he had been a second-lieutenant in the Royal Artillery. He too had flown with 50 Squadron, completing his tour at the end of 1941 to win his DFM. After a spell with two Conversion Units he had returned to 50 Squadron in November 1942 and had flown fifty-one trips by the time he was posted to 617.

Gibson's final crewman was front gunner[1] Sergeant (Pilot Officer wef 18 May 1943) G.A. Deering, who had been born in Ireland of Scottish parents. He was educated in England but moved to Canada where he later joined the RCAF in July 1940. He had completed a tour of thirty-five trips.

One of the first men to arrive was Jim Clay of Les Munro's crew. After finishing his tour with 97 Squadron, Les asked him if he would like to come to a newly-formed squadron. Munro told him he was in need of a bomb aimer and that the operation with 617 would probably be hazardous. Jim was unmarried, and had six brothers in the Army, Navy and Merchant Navy, a sister in a war factory, and his father was in the Home Guard. It seemed right to him to carry on flying, so he agreed to join Munro.

On his reaching Scampton very few people had arrived and there was little to do, so they organised themselves a flight office with a blackboard, etc, and under the helpful eye of Chiefy Powell tried to

---

[1] As the Lancasters of 617, converted to carry the Wallis bomb, did not have a mid-upper gun turret, mid-upper gunners flew in the front gun turret.

obtain whatever else they felt necessary.

Messing at Scampton was shared with 57 Squadron and the embryo squadron personnel had to put up with many a crack about their non-operational role at the base. Another crewman to arrive early was Flying Officer Dave Rodger of Joe McCarthy's crew. Dave was a Canadian and Big Joe's tail gunner. He was amazed at the subsequent gathering of ribbon-decorated men in one squadron as many of the flyers wore DFCs or DFMs.

Towards the end of March, the crews so far assembled were addressed by Gibson, as he stood on the bonnet of a car in a hangar. The importance of the address was clear. All the crews, he told them, were picked volunteers. The odds were heavily against them and anyone who wished to withdraw would be allowed to do so willingly, without the slightest question of LMF (Lack of Moral Fibre – RAF parlance to cover many things. There were no withdrawals.

As the ground staff arrived at Scampton, with its grass runway, they were very pleased to find that it was a permanent station. Brick-built buildings in which to live, hangars in which to service the aircraft – to many this was luxury after airfields of tin huts and open fields.

After settling in at Scampton, over thirty of the ground crew were put on a charge for being scruffy and untidy in their dress. When the charges were brought before Gibson, he sent for Flight Sergeant Powell, asking him for a full explanation. Powell replied that because of the atrocious weather conditions, plus the journey to Scampton, the airmen's clothing was in need of replacement. Gibson immediately rang the clothing officer, who at first was not at all helpful, but with a few well chosen words from the new CO, his attitude quickly changed. The result was that a special clothing parade was arranged for the next day, but before this all ground crew were ordered to parade in front of the squadron hangar. Gibson spoke to them, telling them that all those who had been put on a charge could forget about it. However, since all the unserviceable clothing was being replaced, he expected a far higher standard of turnout from them. He hastened to add that he completely understood the conditions some of the men might have been working under on previous airfields, but nevertheless he would personally inspect them in a few days' time and there should be no reasons in future for charges to be brought for untidy dress.

By his actions Gibson showed everyone that he was in complete command and prepared to back his men to the hilt if their cause justified it. Gibson, a supremely confident pilot and leader, had only one aim in mind – to build up his new squadron into a team in order to complete the operation successfully.

Gibson then raised the question of security, saying: 'You will no doubt see many strange happenings and exercises, but whatever you see or hear you must not talk about, least of all in the local pubs when you go for a drink.' His final words to the assembled men were addressed to the airframe and engine mechanics. 'Should any of the aircrew wish to go on a test flight and invite you to go along, for goodness sake go, and show them that you have confidence in your own workmanship. All I ask is that, if you do go flying, just make sure that someone knows you have gone. In this way you have no need to put your name in the flight authorisation book.'

Meanwhile, on 24th March, Mutt Summers had driven down to Burnhill, to Wallis' drawing office. With him went Gibson, who was introduced to Wallis for the first time. Wallis asked Gibson if he had yet been told the target. To this Gibson replied: 'Not the slightest idea.' This made it awkward for Wallis, as only the names on a list specially drawn up, a copy of which he had, were to know the actual target at this time, and Gibson was not on the list! After the meeting, Gibson returned to Scampton, fully informed of the workings of the bomb and how 617 Squadron were to employ it, but still had no idea of the actual target. Naturally the whole show was highly secret and the following codenames had been selected:

Highball – The Special Weapon, smaller version of the dams bomb.

Upkeep – The Special Weapon, as used on the attack on the dams.

Chastise – The Operation against the dams.

By the end of March 1943, fifteen of Gibson's pilots had arrived and five more were on their way. Most of his ground personnel had joined and been briefed, and about to join were his Squadron

(*Left*) 'Tammy' Simpson – Blida, North Africa. (*Right*) Flight Sergeant Charles Franklin DFM.

(*Left to Right*) Bill Howarth, John Pulford, Percy Pigeon, Harvey Weeks.

Adjutant, Flight Lieutenant Harry Humphries, coming in from 50 Squadron, Engineering Officer, Squadron Leader Clifford Caple, currently at Group HQ, Armament Officer, Flight Lieutenant Henry Watson, from 83 Squadron, and finally the Medical Officer, Flying Officer Malcolm Arthurton.

Sir Arthur Harris had also been requested to supply more than just three Lancaster bombers, the new aircraft being due to arrive shortly. Rumours of why the squadron had been formed with such an array of variously experienced pilots and aircrew, and with such a highly decorated 'boss' were rife. These rumours ranged from a possible attack on the *Tirpitz* (the 'official' cover plan) to a special mission to kidnap Hitler.

Gibson could only speculate about the target. Training under his keen direction was about to begin, but training for what? The answer to that question would have to wait.

CHAPTER FOUR

Training and Build Up

The new 617 Squadron began training on 31st March. This mainly consisted of low cross-country flights, getting used to flying lower than any of them had been accustomed to, and also to navigating at zero feet.

The entry in Flying Officer Robert Urquhart's log-book, navigator to Squadron Leader Maudsley, for 31st March reads: 'X-country 500 feet; low level bombing, speed 240 mph – 100 feet.' The exercise was carried out by twenty crews on standard Lancaster bombers, prior to the arrival of the special Lancasters. At the end of the first week, some 26 cross country flights and 240 bombs had been dropped (conventional type), the average error margin being only 40 yards.

All these flights were flown in daylight. Night flying was expected to start around 10th April which was estimated to be the right time for suitable moonlight conditions. Sergeant Tom Jaye, Lewis Burpee's navigator entered in his log-book for 1st April, 'Low-level X-country and bombing.'; an entry that was repeated for the 3rd and 4th, adding 'air firing' on the 4th, when the air gunners practised gunnery from low level. The area chosen for these flights were over and around lakes in North Wales.

One problem of flying over water at night was how to judge the correct height of the aircraft above the surface. Ben Lockspeiser (knighted in 1946), Director of Scientific Research at the MAP, was approached to try and find a solution to this critical problem. His answer was to place two spotlights on the Lancaster, one under the tail and one in the nose. Set at pre-determined angles they cast a spotlight on the surface of the water. As the aircraft reduced height, so the two spots of light met, and when they did so then the aeroplane was at the desired height. The idea was first used in

WW1 but it met other snags and was shelved. In WW2 it was regenerated for use in anti-submarine warfare by the Royal Aircraft Establishment. However, it proved to be a failure as the choppiness of the sea, made it impossible to identify when the lights met. But over smooth water, such as lakes, there was a good chance that it would work. With this in mind, Lockspeiser visited Cochrane and convinced him that the idea was worth a try. He agreed and a few days later a Lancaster was fitted with the lights at Farnborough, then returned to the squadron for trials over Lake Windermere. The trials were a success and the problem of maintaining the right height was solved.

In April 1943, Corporal Maurice Statham was an MT driver with 5 Group HQ, being the personal driver for the AOC. He recalls that on 5th April Station Officer Carol Durant, personal secretary to Cochrane, detailed him to drive to RAF Scampton, collect Wing Commander Gibson and drive him back to Grantham. It was at this meeting on the 5th, that Gibson was finally told what his targets would be. Corporal Statham also recalls the efficiency and control that the AOC had in organising the whole show, which was probably beyond the imagination of most people.

In consequence of his knowing the targets, Gibson was told to practise low level flying over various selected reservoirs in preparation for the big-day. The entry in Gibson's log-book for 4th April shows: 'Lake near Sheffield' (This was probably the Derwent), and for the 5th: 'Scotland X-country, Lakes.'

Pilot Officer Lance Howard, Sergeant Townsend's Australian navigator, remembers most the hard work involved flying in at 60 feet, unable to use the automatic pilot and wondering if Bill Townsend would be able to stand up to the bouncing about that he was subjected to up front in the pilot's seat. This was due to the warmer spring weather, the warm up-currents giving the aircraft a great shaking. But he says that Bill was a natural pilot and did a magnificent job. Cecil could sympathise with his pilot as he himself had started his training as a pilot but as he would persist in landing an Anson twelve feet from the ground, he was remustered as a navigator, serving with 49 Squadron in 1942-43.

Another who remembers this period is Leading Aircraftman

Keith 'Lofty' Stretch. With all the training flights going on, many of the ground crew were able to hitch rides on them, the aircrew always willing to take them, except when flying an exercise of a more secret nature. Lofty recalls, 'It was a great thrill hedge-hopping in a Lanc piloted by Flying Officer Les Knight; his skill and calm is something I will never forget.'

All through April training continued and damage to trees was often reported as the Lancasters were chopping off the tops as they scraped over them at low level. However, the low flying and bombing was improving greatly as was the navigation.

As 617 was formed for a special mission it was acknowledged from the outset that good air to air radio transmission range at heights from 50 to 1,000 feet was a necessity. The aeroplanes initially supplied had the TR 1196 radio transmitter fitted and tests were made to ascertain whether the set could meet the requirements of the operation. The first test on 28th March was in daylight, carried out at a height of 2,600 feet on 5005 KCS R/T, and communication at 40 miles range was obtained. Further daylight tests took place on 7th April between two aircraft from 500 feet, up to 15,000 feet with an average separation range of 30 miles.

In one of the aircraft were Guy Gibson and 5 Group's Chief Signals Officer, Wing Commander W.E. Dunn, who had a wealth of experience in signals behind him. The result of this test was quite good, but to be doubly sure a night test was flown. On this the results were hopeless, because of continental waves and the background noise that occurs at night. The problem was left with Wally Dunn for the time being.

During the week ending 15th April, a week of good weather, many of the teething troubles in training were ironed out. Flight Sergeant Lovell had arrived from 57 Squadron but was replaced by Flight Sergeant Divall from the same unit. Divall was naturally behind the others in training but was expected to catch up by the end of the week, as was Pilot Officer Ottley who did not join 617 until 6th April. Young, Anderson and Wilson also had only just or were about to join but they too would have time to catch up. Time in the air for Dinghy Young was reduced, however, when Gibson went on an attachment to RAF Manston where he was learning more about the weapon his squadron would be dropping on the dams. As senior flight commander, he had to 'look after the shop!'

There was also a shortage of aeroplanes so Flight Sergeant Brown was sent to Fulbeck for a short Bomb Aiming Training Course for three days. During April, the low level flying height was reduced to an amazing 20 feet while the weather held good, although some cloud over hills restricted training over the eastern counties. Then on the 14th, the weather allowed a training flight to the north of Scotland. In early April the number of exercises flown and bombs dropped, with error margin rate, was:

Date	No of Exercises	Bombs dropped	Error Margin in yards
9th April	6	92	42
10th April	10	84	42
11th April	12	136	43
12th April	8	63	40
13th April	7	59	48
14th April	7	60	35
15th April	6	55	37

Lofty Stretch remembers the 15th. The ground crews had prepared the aircraft for take-off and as the Lancasters began to move off to the take-off point, so the ground people began walking back to the flight huts. They had seen Lancasters take off enough times not to watch particularly on this day. But Lofty remembers that he suddenly heard the Lancs making a louder noise than usual and looking back saw not just one bomber taking off in turn but three in formation, followed by the other section three. They took the whole length of the grass runway before lifting off, then they skimmed the hedges before making a slow bank to the left, disappearing from view for a few moments. When next seen they had all formed up and were flying over the field at about 200 feet, by which time it was dark. The ground crews were of the opinion that the pilots had gone quite mad!

With all this low flying going on it was not long before people living in the various chosen areas began writing to complain of the noise, but there was a war on, and training carried on.

On 11th April, Gibson and the squadron's Bombing Leader, Flight Lieutenant R.C. Hay DFC, of Mick Martin's crew, had flown down to Manston in a Magister, to witness the first test of the Upkeep bomb being dropped. Bob Hay, from Gawlor, Australia, had served with Martin in 455 and 50 Squadrons, receiving the DFC at the end of his tour, Hay having completed thirty-four trips.

The test was at Reculver Bay, near Margate. The drop was to be made at a speed of 270 mph and the pilot, Sam Browne, was instructed by Wallis to give the bomb back-spin of 300 rpm and to let it go when level at 150 feet. Dead on time two aircraft appeared, one carrying a cine-camera. As they neared the white marker buoys, placed some yards off shore, Wallis realised the aircraft carrying the bomb was too high. The bomb came away from the aircraft but instead of bouncing on the water it sank out of sight immediately. Another test was carried out and this time the bomb was dropped at a much lower height. This time, instead of sinking it disintegrated in a shower of wooden staves, steel bands and bolts; the heavy steel cylinder burst out with such force that one wooden segment smashed into the Lancaster's elevator as it passed over, nearly causing disaster to the bomber. The elevator jammed, causing Mutt Summers, the pilot on this second test, a serious problem in maintaining height both then and later when landing.

Training on the squadron continued through the week 16th to 22nd April, although there was a slight decrease in flying due to aircraft being stood down for their periodic inspections. Lack of experience on the part of some of the crews had held up the training programme slightly. Map-reading at night at heights of 150 feet over water, and how to carry out the special attack approach were the important items now. The map reading fell mainly to the bomb aimer as he lay in the nose of the Lancaster. The track was previously calculated and drawn from forecast wind speed and direction. The two gunners would take drifts from time to time from the turrets, also the bomb aimer through his bomb sight. This allowed the navigator to adjust course as necessary. The bomb aimer endeavoured to give pin-point positions mainly by visual sightings over land, and drifts calculated from flame floats dropped over water. A shortage of these was overcome by the SASO at 5 Group, who made 500 available to the squadron so they could operate through the moon period and on to the end of the month. Another job that fell to the bomb aimer was to assist the pilot in watching ahead when flying a low level, keeping a sharp look-out for power lines, trees, tall chimneys, etc. Flying fast and low, a pilot's momentary glance at his instruments would mean his eyes would be off the darkened landscape for perhaps a quarter of a mile or so, in which time all manner of things could suddenly loom ahead of them. Night-flying cross-countries had been flown regularly and all crews had carried out one special night cross-country over a route very similar to the actual conditions to the target.

The weather on the whole proved good, 617 being able to fly at night on five out of six nights in mid-April. Gibson's log-book entry for 16th reads: 'Cornish X-country at low level with dummy attacks on lakes.' During the week all navigators and bomb aimers attended a film on map reading, it being agreed the lesson was necessary for systematic and methodical work while in the air.

Day bombing from 100 feet by day and 150 feet by night continued at Wainfleet Sands Bombing Range. A speed of 240 mph had now been stressed by Wallis, as the pre-set speed at release. The number of bombs carried by each aircraft had been reduced to six, there being no benefit from dropping a larger number. During this mid-April period the following exercises were flown.

Date	No of Exercises	Bombs dropped	Error Margin in yards
16th April	6	32	39
17th April	4	24	43
18th April	3	18	43
19th April	nil		
20th April	6	42	53
21st April	6	39	55
22nd April	3	24	43

After a month of cross-country flying all the crews could navigate from pin-point to pin-point at low level over water at 150 feet. The height at which Wallis now knew the bomb had to be dropped at was exactly 60 feet, and this height was introduced to the crews on 26th April, together with a new air speed of 210 mph. Although they did not know it, the date of the raid was now just three weeks away.

Sixty feet in a fast flying, four-engined Avro Lancaster, is extremely low and extremely dangerous. The operational squadron aircraft had their spotlights fitted in such a position that the degree of travel of the spots at 60 feet was roughly the same as that fitted in the training aircraft for use at 150 feet. Crews were briefed in connection with the spotlight training, that if the aircraft descended too low it must be immediately pulled up sharply to a safer altitude before going down again for another try. The positioning of the spotlight was decided by three factors: (i) to simulate training at 150 feet; (ii) to prevent oil smearing over the rear spotlight, and (iii) to shield the light as much as possible. There was a decrease in flying hours because only two of the Lancs had been fitted with Synthetic Flying Equipment, something that Gibson was trying to sort out.

This equipment, also known as the Two Stage Blue Day-Night Flying System, was invented by two brothers early in the war, Squadron Leader Arthur Wood and Squadron Leader Charles Wood. Arthur had been a fighter pilot in the First War, and both he and his brother saw service with the Technical Branch in WW2. Charles had a photographic business before the war and from 1940 to 1946, when he received the MBE, he worked entirely on Day-Time flying systems.

The training for the dams raid had many problems but one major one was how to complete its training for an operation in full moon conditions, when the full moon periods came monthly. Over the whole period of 617's training, there were only a few nights of ideal brightness. The Wood brothers solved problems of this nature by allowing an aircraft to fly in daylight while simulating moonlight conditions. This was achieved with the use of light-absorption filters – or colour filters – that could either cut off or produce different colours of the spectrum. A blue filter, for example, allowed the eye to see only the blue and green wavelengths of light, cutting off the yellow and red, whereas its complementary amber filter cuts off the blue and the green, allowing only vision of the yellow and red. The two together produced a complete blackout. In the Day-Night Flying System, all the windows of an aeroplane's cockpit were covered with a blue filter component, whilst pilot, navigator and bomb aimer wore goggles with amber-coloured glass. The effect was that all three men saw everything inside the aircraft – controls, instruments, maps, etc, all in a subdued reddish light, exactly as one would expect to have at night in operational flying, whilst their vision through the blue windows simulated a dark or moonlight night. Gibson approached Charles Wood and eventually he had five aircraft fitted out with the system for use by his squadron. The first arrived on 15th April; the other four arrived by the end of the month.

With these aircraft, crews were able to train for some six weeks on intensive low level training at any time of the day, flying as though in perfect moon conditions. Much of this training with the system was carried out over the water at the Derwent Valley Reservoir in Derbyshire, which in many ways resembles the Möhne Dam in Germany. From 5th May training started on a reservoir four miles south of Uppingham.

One person who remembers the fitting of the system to 617's aeroplanes was Mr White, of Huddersfield, Yorkshire, employed by A.V. Roe Ltd as an instrument fitter. During April 1943 he was called into the office of Phil Lightfoot who was in charge of company activities at RAF Waddington, and told that the work he was currently doing would have to be halted, and room made for a Lancaster which was arriving from Scampton. As soon as it arrived he was to carry out urgent modifications involving a parcel that had already arrived, and which lay on the floor of the hangar, and blue perspex that would arrive the next day. He was to have three men to help him. When the Lancaster arrived it caused quite a sensation with the omission of its bomb doors, special release gear (the squadron were now dropping cylinder shaped 'bombs') and what appeared to be two extra landing lights fore and aft. It all seemed highly odd, as did the job they were about to start on.

This job entailed covering the windscreen, canopy and side windows of the cockpit, with removable perspex panels. Although precise measurements were shown on a drawing, brown paper was used to make templates from which the panels were cut out to size. This was necessary owing to slight variations during fabrication of the cockpit could not be avoided, thus every Lancaster's window panels could vary slightly from one to the next. The male half of a dot fastener was affixed to the cut out, window frames and panels were given a trial fit and each fastener point marked off. The completed panels were taken to the WAAF workshop, where black linen hems were machined on, then the female half of the fasteners were fitted to the hems. Then the panels were fastened into position and after inspection, the Lancaster was taken out onto the tarmac for a vibration test with engines roaring at full throttle.

On the morning the job was completed Phil Lightfoot told Mr White that a Wing Commander Gibson would be arriving to inspect the job and he was to make any alterations the Wing Commander might require. When Gibson arrived, they climbed into the Lancaster, exchanging a few formal words. Once inside, White decided not to make any comments but to let the Wing Commander make a thorough inspection and decide himself if or not he was happy. Gibson sat in the pilot's seat and tried on one or two pairs of goggles, at the same time examining the instrument panel in front of him. When he seemed finished, White asked him if he was satisfied. 'Yes,' replied Gibson, 'everything seems to be OK.'

Later that day Lightfoot ordered two more aircraft modified, but they would not be coming from Scampton – none could be spared from the training programme.

Security at Waddington was often quite strict and so it was on this occasion, but by now, at RAF Scampton, it became much stricter than normal. Each new arrival was issued with a pass which included fingerprints and a photograph of the person. All personnel were instructed not to refer to their work or the special aircraft on or off the station, and if found doing so they would be immediately put under arrest.

The Waddington conversions took ten to fourteen days. What seemed quite a simple job took much longer, having to work over or around the pilot's seat, the control column and all the other knobs and tits in the cockpit. The main concern of Mr White and his men was to try and avoid bulges and tension in the perspex, otherwise vibrations on take-off would spring the fasteners apart.

Meanwhile, two screens had been erected at Wainfleet Bombing

Range, 750 feet apart, for simulation of the actual attack, but a gale on the 26th blew them away. More were built and kept at Wainfleet until the location was changed to Uppingham. Exercises continued:

Date	No of Exercises	Bombs dropped	Margin Error in yards
23rd April	nil		
24th April	19	165	47
25th April	5	54	26
26th April	nil		
27th April	7	65	34

In order to judge precisely the distance to drop the bomb from the dam, a special range finder sight was devised, constructed and fitted to the aircraft. The idea originated from Wing Commander Leslie Dann, a pre-war regular who had joined the Technical Branch in 1940. (Later Air Commodore CBE, died 27th February 1965 having retired from the RAF in 1957.) Dann's idea was simple. Construct a triangle of plywood with a peep-hole (or backsight), and fix two upright nails, one each, to the frontmost corners. By looking through the peep-hole the bomb aimer watched the two towers on the dam wall as the aircraft flew towards them. When both towers corresponded to the two nails, the aircraft was at the point of release.

Another problem arose with all the low flying exercises; air sickness. Here the Squadron MO, Flying Officer Arthurton, tried to help out, flying in Maudsley's Lancaster ED906 on 25th April when they practised low flying over Derwent water. In his own log-book, Maudsley wrote: 'Low flying, weather bumpy – everybody airsick after half an hour. Total flying time one hour.'

Although the low flying had more or less been mastered, the recent lowering of the height to 60 feet had caused several narrow scrapes with much damage being caused to aircraft. One damaged was that of Squadron Leader Maudsley who came back with his Lancaster so battered it was surprising that he ever got back at all. Airframe fitters spent many hours replacing damaged panels. One other diversion remembered by Jim Heveron was one crew coming back from a flight to Scotland, having landed en-route on the Isle of Mull. There they had picked up a few seagulls' eggs but after eating them they decided that it had not been worth the effort!

While the squadron was becoming proficient, Barnes Wallis was still testing his bomb. Time was running out. The high water period on the Möhne with a moonlight night was just three to four weeks away!

On 28th April a Highball bomb was dropped from a Mosquito at a height of 130 feet at 350 mph into a head wind of 15 mph, revs: 700 per minute. It made four or five bounces, travelled 1,000 yards, and although slightly dented on impact suffered no major damage. Another was dropped the next day, armoured with steel plates (5/32"); height 60 feet, speed 370 mph into a 5 mile-an-hour head wind at 920 rpm. This bomb behaved well, making four or five bounces, travelling 1,000 to 1,200 yards and suffering no damage at all.

On the same day one cylinder Upkeep bomb was tested. It was encased in ash wood. Dropped at 60 feet, at 258 mph into a 5 mile head wind, revolving at 500 revolution per minute, it bounced four or five times, covering 600-700 yards. The bomb travelled straight at first but then deviated slightly to starboard[1] towards the end of its run. The sea was calm for the drop with only a slight swell of about one foot. The sub-committee considered that on all counts, Operation Chastise should be authorised immediately. If the decision was not made shortly the chance of delivering the attack would be lost until 1944.

The squadron's operationally modified Lancasters began arriving at the end of April, and assigned to squadron pilots:

ED887 to Young on 22nd April	ED909 to Martin on 23rd April
ED865 to Burpee on 22nd April	ED886 to Townsend on 23rd April
ED864 to Astell on 22nd April	ED906 to Maltby on 23rd April
	ED910 to Ottley on 28th April

Over the first week of May a good deal of training with the spotlights took place on all types of water, making dummy attacks

[1] It had a tendency not to run dead straight and with the rough water that developed on the Möhne dam during the attack, the bounces of the bomb could have been deflected or shortened.

over the Eyebrook Reservoir, until all crews were confident in their use. Meanwhile a further five Lancaster aircraft arrived: ED932 for Gibson, ED925 to Hopgood, ED921 to Munro, ED918 to Brown, ED929 to Shannon and ED924 for Anderson.

The operational aircraft were used at once, log-book entries confirming intensive testing and exercises flown over the first week in May. The crews took great pains to get their aeroplanes ready and were fitting any small gadgets they felt were necessary. One very good modification was the introduction of an additional sensitive altimeter, which took the place of the visual loop direction indicator. This enabled the pilot to fly at night without having to lower his eyes from the horizon.

The bombs arrived. In all fifty-eight were delivered, thirty-seven to Scampton and nineteen to Manston. As yet they were without their explosive content. Also delivered to Scampton was one 10-ton Coles crane, six modified bomb trolleys, two mobile gantries, three sets of lifting tackle and twenty winches.

Strict instruction were given that no attempt should be made to fit the bombs to the Lancaster aircraft. Pilot Officer Watson, the Armament Officer, had been attached to Manston for a month while the trials were in progress, returned on 1st May and would be responsible for bombing up when the date was set and the time came. In the meantime, the bombs were kept securely locked away and out of sight – there was enough speculation going on as it was.

Most of this came about when the new Lancasters arrived, the absence of a top turret, no bomb doors and with extensively altered bomb bays, was hard to disguise. The alteration in the bomb bays was the two triangular support arms for the bomb and the VSG motor for use when spinning the huge cylinder. This motor was very compact, 15 to 18 inches in height and hydraulically driven by a pump from one of the inboard engines, the same sort of system that operated the flaps and undercarriage.

All leave was stopped from mid-day on 7th May. On arrival back at base all aircraft began to fly low over the airfield in order to check the calebration of the spotlights. Since training had commenced, a total of 31 exercises had been flown of a specific rather than general nature, involving 168 drops, of which 52 had been deemed successful. The low per-cent of the success was due to

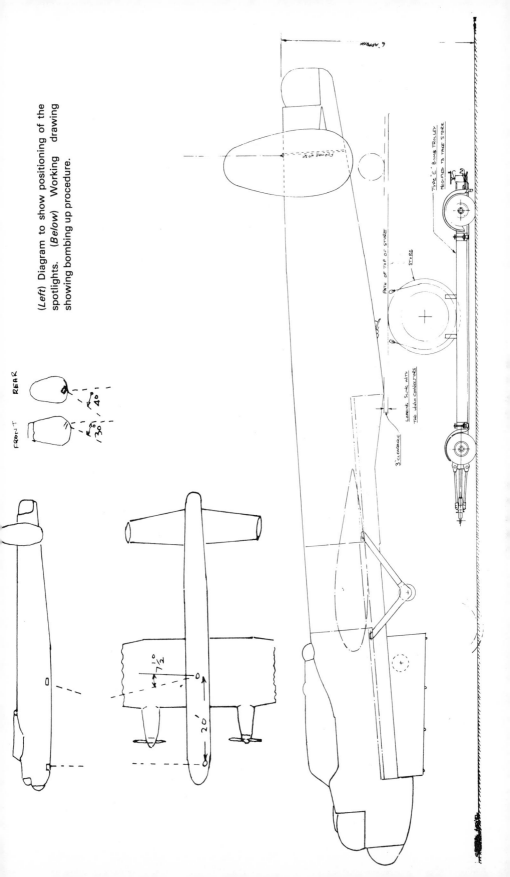

(*Left*) Diagram to show positioning of the spotlights. (*Below*) Working drawing showing bombing up procedure.

the fact that although crews were getting 100% success by day, night exercises had only been directed against small flags at Wainfleet and these had been hard to see. However, when the boards went up in place of the flags, results were expected to increase to around 90% successful.

Most of the bomb aimers had modified their aircraft, with the help of Flying Officer Caple – known to everyone as Capable Caple – the Engineering Officer. To give greater comfort for both map reading and bomb dropping, the chest rest on which he usually laid had been lowered and repositioned to give more freedom of movement. Additionally stirrups had been fitted to the front gun turret so as to keep the gunner's legs out of the bomb aimer's way. Irrespective of their normal position, the smallest of the two air-gunners in each crew occupied the front turret for the raid. There was also another variation on the bomb sight/range finder. This was a pre-calculated piece of string attached to the side nuts of the clear vision panel in the nose and this was used by some bomb aimers rather than the hand-held wooden triangle sight.

Discussions had been held with all the gunners on deflection and air to ground firing experiments, to find the best mix of ammunition to be used. It was finally decided that to give maximum 'scare effect' every round should be tracer.

By now eighteen of the twenty operational Lancasters had arrived on the squadron, and had been serviced and prepared for the attack.

Wing Commander Wally Dunn, left with the problem of inter-aircraft communication, had flown another training flight with the squadron on 4th May, again with Gibson. With communication imperative on the night, Dunn felt that VHF (Very High Frequency) radios must be used. Previous flight tests at night still ran into interference problems. The test flown on 4th May, in daylight, over a distance of zero to 60 miles to another aircraft, flown by Dinghy Young, proved satisfactory but Dunn insisted that another night flight test should be made, around mid-night to 1 am – the proposed time of the raid. They flew this the same night but it was hopeless, static blotting out everything. Dunn told Gibson that he would enlist the aid of the RAE, and Wing Commander

Lancaster ED 817-C
(Not used on the raid)

The Wallis Bomb

Lancaster ED 825-C
(Not used on the raid)

Allerston of HQ Bomber Command arranged for Flight Lieutenant Bone to assist.

The successful use of TR 1196 depended on the reception of strong 'wanted' signal in order for the automatic volume control should operate and reduce interference. This 'wanted' signal was not forthcoming at ranges in excess of thirty miles. Bone suggested two alternatives, one of which was to install VHF transmitters – type TR 1143 (100-124 MC/S) as used in day fighter aircraft.

Wally Dunn explained the position to the AOC and to the Deputy Chief Signals Officer at Group HQ on 6th May. As a result VHF radios were immediately ordered for the Lancasters of 617 Squadron. By 5.30 pm on Sunday 9th May, all eighteen available Lancasters had been fitted, which indicates the drive and enthusiasm with which this seemingly impossible task was handled. The RAE Radio Department had had to 'prototype' the installation and No 26 (Signals) Group provided the manpower. By Friday the 7th the equipment had been delivered to Scampton and an RAE representative, plus an officer and twenty-five men from the Signals Group had arrived.

During Friday, the man from the RAE was occupied with the prototype installation in co-operation with the squadron specialists and the first installation completed and ground tested by the evening. No 26 Group party was detailed into parties, and by working in shifts completed the whole job by Sunday afternoon.

A flight test between two aircraft was set up for Sunday evening with the flight commanders as pilots, the Group's Chief Signals Officer, and the RAE as observers. It was arranged that Squadron Leader Maudsley should fly a course of 180° at 500 feet and that Squadron Leader Young should circle the aerodrome at 500 feet. Communication was then to be made and held to nearly extreme range, when Young would make flat turns around Scampton and indicate the course over 10° by Radio Transmitter, so that the observer flying with Maudsley, could make an estimation of the Polar Diagram to be expected.

During the flights, good communication was maintained to about fifty miles and the Polar Diagram test did not indicate any blind spots. Various adjustments were made after this test and by 11th May all modifications had been completed on all aircraft. Also

nne Dam before the raid.
e the power station on
right.

Sorpe Dam today

Ennepe Dam today.

on this day Gibson and Martin tested a modified aircraft on an Upkeep drop from 50 feet.[1]

The Squadron pilots practised the use of these VHF radios on the ground, being wired up in the crew room to enable pilots to become word perfect in their R/T drill. An inter-communication system including twenty sockets wired in a common circuit, was set up, running through both flight commander's offices and around the crew room. Each pilot plugged into a socket and by means of an A1134A Amplifier, the entire R/T procedure was practised until they knew exactly what was expected to be done on each order received. This scheme saved many flying hours, which was just as well, for the moon period was almost upon them.

The final form of Upkeep had now been decided upon, and twenty should be ready within three days. Trial spins had been carried out at 500 rpm and it took thirty to forty minutes for the bomb to stop rotating. Airframe mechanic Payne remembers his first sighting of the bomb: 'If you imagine the front wheel of a steam roller, its size and shape, then you have a fair idea of what the bomb looked like.'

On the 14th a full-scale rehearsal was flown over the Eyebrook and Colchester Reservoirs. The Squadron MO was again flying in Maudsley's aircraft and he entered in his log book: 'Experiment of taking Chloretone for airsickness; result – no nausea or airsickness among the crew. Flying time four hours.' Although Arthurton's log shows this to be in Lancaster ED906, it was in fact ED937 'Z' which was the last modified aircraft to arrive on the squadron.

Guy Gibson took off for the rehearsal at 10 pm; with him went Group Captain Whitworth the Station Commander, flying as observer, the third aircraft being flown by Dave Shannon. The rehearsal proved a complete success. The way was now clear for the raid to take place.

A letter dated 10th May, 1943, addressed to Sir Arthur Harris and Bomber Command at High Wycombe (HQ Bomber Command),

[1]At this height the water spout came up and hit Martin's elevators causing some damage and it was then decided to raise the height to 60 feet.

from Air Commodore S. Bufton, Director of Bomber Operations, contained the following:

I am directed to inform you that in order to avoid specific mention of the Möhne, Eder and Sorpe dams, in future correspondence and discussion, it has been decided that these dams should be called by the following:

Möhne Dam	–	Objective 'X'
Eder Dam	–	Objective 'Y'
Sorpe Dam	–	Objective 'Z'

CHAPTER FIVE

Briefing

On Saturday, 15th May, all pilots and navigators were finally informed of the target. There was a great sigh of relief after weeks of speculation, and they felt they had at least a sporting chance of survival, for the dams might not be as heavily defended as the *Tirpitz* was known to be. They were told not to tell the other members of their crews, and despite much questioning the secret was kept for a further twenty-four hours.

The ground crews were also wondering, after all the training, when the big day would come. At mid-day on the 15th, the Senior Air Staff Officer instructed Wing Commander Dunn to read the draft order, and from it devise a signal instruction.

Code words were devised to cover every eventuality. Some had already been used during the practice period, and although such words as 'Dinghy' and 'Danger' had appeared unsuitable at first, it was considered unwise to alter them at this late stage. A special code was devised to enable each aircraft to indicate:

1. That the special weapon had been released correctly,
2. Where, in relation to the target, the weapon had fallen,
3. The condition of the target.

At ten o'clock on the morning of 16th May the Chief Staff Officer was informed that the operation was 'on' for that night. He proceeded at once to RAF Scampton and had produced sufficient copies of the codes for each crew. All wireless operators were carefully instructed in W/T procedure, actual specimen messages being transmitted on a buzzer circuit. The VHF was to be the primary method of control and a frequency set up on Button 'A' was to be employed. The use of Button 'C' would serve as a reserve frequency by the second group of aircraft. The code-word 'Codfish' went by W/T was the signal to bring the reserve frequency into force. From take-off to zero hour, all aircraft were to maintain a listening watch on the TR 1196 R/T set using Button 'D' at 0300 E. The VHF set was to be used ten minutes before the target was reached and used until one hour after the completion of the attack, when operators were to revert to the 1196. All R/T was to be in simple language and conform to the system used during ground practice. In the event of the leader's R/T becoming faulty, he will inform either aircraft No 2 or No 4 to transmit the codeword 'Deafness' by W/T twice.

The wireless operators were also briefed to maintain a continuous listening watch on 4090 KCS throughout the operation except when passing the 'Operation Completed' signal on 3680 KCS. In the event of a complete failure of control on VHF the leader of the first wave was to transmit by W/T the codeword 'Mermaid' which meant jamming. On all W/T the following codewords were to be used:

Pranger	–	Attack target 'X'
Nigger	–	Target 'X' breached, divert to target 'Y'
Dinghy	–	Target 'Y' breached, divert to target 'Z'
Danger	–	Attack target D
Edward	–	Attack target E
Fraser	–	Attack target F
Mason	–	All aircraft return to base
Apple	–	First wave listen out on Button B
Codfish	–	Jamming on button A, change to button C
Mermaid	–	Jamming on all R/T control by W/T
Tulip	–	No 2 take over control at target 'X'
Cracking	–	No 4 take over control at target 'Y'
Gilbert	–	Attack last resort targets as detailed
Goner	–	Bomb released

Group HQ would repeat the whole message twice using full power. Should W/T control be used, each aircraft was to call the leader of the group by W/T as soon as it arrived over the target. Instructions would then be given by the leader to attack an appropriate target.

Early on the morning of the 16th, all crews were ordered to the huge airmen's dining room, while outside stood three guards. Including Gibson's, nineteen crews had been selected for the attack, therefore 133 crew members plus the Station Commander, weather, signals and station officers, together with the AOC, Ralph Cochrane, Barnes Wallis and Wing Commander Dunn, assembled in the large hall. The two unlucky crews who would not be going were those of Sergeant Divall and Flight Lieutenant Wilson. On a platform sat Cochrane, Whitworth, Wallis, Dunn, Gibson and the met. officer. Cochrane stood as the noise abated.

'Bomber Command,' he began, 'has been delivering the bludgeon blow on Hitler; you have been selected to give the rapier thrust which will shorten the war, if it is successful.'

When he had finished his address, Gibson stood up, and pulled apart two curtains behind which were three large photographs of the dams, also a large map showing the routes to them and table-top models. These latter had been made by the RAF staff at Medmenham, the home of Photographic Intelligence. Dave Rodger remembers his pilot. Joe McCarthy, saying, 'It sure looks big to breach.'

Gibson's briefing outlined the plan for the squadron to fly from Scampton in three groups or waves, to attack the Möhne, Eder and Sorpe dams, in moonlight at low level. The three attack groups were:

First group: nine Lancasters in three sections spaced at ten minute intervals, each consisting of three aircraft led by Gibson. They would take the southern route to the target area and attack target 'X' – the Möhne. The attack would be continued until the dam had been clearly breached. It was estimated that this might require three 'effective' attacks. When this had been achieved the leader would divert the remainder of the group to target 'Y' – the Eder, where similar tactics would be employed. Should 'X' and 'Y' be breached, any remaining aircraft of the first group still with its bomb would fly to the Sorpe – target 'Z'.

Second group: five Lancasters led by Squadron Leader McCarthy and manned by those crews who were to take the northern route to the target but would would cross the enemy coast at precisely the same time as the leading three Lancasters of the first group, but at a different point. This second group would attack the Sorpe, and by flying in from the north would act as a diversionary force for the first group.

Third and final group consisting of the remaining six Lancasters, led by Flight Sergeant Townsend, would form an airborne reserve under the control of Group HQ. They would fly the southern route to the target but their take off time would be such that they could be recalled before crossing the enemy coast if the first and second groups had successfully breached all three of the targets.

The Lancasters would all fly in open formation but height must not exceed 1,500 feet over England. On leaving the English coast, the pilots would descend to low level and set their altimeters to 60 feet over the sea by using their spotlights for calibration.

The crews were told that on arriving at a point ten miles from their target, the leader of each section would climb to 1,000 feet. While the other aircraft would listen out on VHF, each would call the leader of the group of VHF on arrival at the target and spinning of the weapon would be started ten minutes before each aircraft attacked. The leader was to attack first and would then control the following attacks by all other aircraft of the first group, using the laid-down signals procedure.

Number two of the leading section of the first group – Dinghy Young – was assigned to act as deputy leader of the first group during the attack on the Möhne should the leader fall out, the deputy would take over leadership and number seven – Henry Maudsley – would take over as the new deputy leader.

All other aircraft were ordered to return to base after completing their attack. When the Möhne was seen to be destroyed beyond all possible doubt the leader would order the remainder of the first group still with bombs on to the Eder by W/T and VHF, where similar tactics would be employed against the dam. This scheme was to be repeated against the other dams. It was imperative for the attacks on the Möhne and Eder dams, that the special range finders be used.

The second group would go straight for the Sorpe, but was not controlled by any leader but would attack independently. The bomb would not be spun for this target, it being impossible to make a bouncing bomb approach. The bombs had to be dropped by flying over the length of the dam, ie from side to side, and aimed to hit the water just short of the centre point of the dam, about twenty feet out from the edge of the dam wall. Attacks on the Sorpe had to be made from as low as possible and at a speed of 180 mph. Aircraft would return to base independently.

The final reserve group under the control of Group HQ would fly towards the Möhne. Orders would be passed to these aircraft by Group on a special radio frequency, hopefully before they reached the enemy coast, if the dams had by that time been smashed. Otherwise they would be directed to an appropriate target. Failing to receive a message from Group, the aircraft would proceed to the Möhne or Eder and as a last resort the Sorpe to attack any or all that had not been breached, using attacks as detailed to the first two groups, i.e. either by bouncing bomb at the first two or a straight crossways drop at the Sorpe. There would be no control other than visual or wireless signals.

During the actual attack, the pilot would be responsible for line, the navigator for height, the bomb aimer for range and drop, the flight engineer for speed. For all aircraft the interval between each attack was put at at least three minutes in order to allow the preceding aircraft's bomb to explode and blown water to subside. On all targets except the Sorpe, the wireless operator in each aircraft must fire a red Very cartridge immediately over the dam during the attack to advise the attack leader that the aircraft had attacked and that their bomb had been released. All aircraft must fly left-hand circuits in each target area, keeping as low as possible while waiting their turn to attack.

The time of attack on each target by each group was not important to within a few minutes. The time of crossing the enemy coast was, however, all important! The whole essence of the operation was surprise and to avoid bringing enemy defences to an unnecessary degree of alertness, diversionary raids were being laid on and timed accordingly. HQ Bomber Command were arranging certain diversions so that the first enemy RDF (radar), or other warning, of the diversionary attacks occurred twenty minutes after the leading section of the first group crossed the coast, and then for a period of one hour preceding the third group's arrival. Fifteen minutes after the third group crossed the enemy coast further diversions would be made at maximum strength and would continue if possible until the third group was clear of the enemy coast of the return journey. At the briefing there were no less than forty-one sets of instructions for the crews to read, learn and digest.

Their routes out were carefully studied before the operation began and the outstanding features, obstructions and pinpoints noted, particularly water pinpoints. This had been why the navigators had been told of the target the previous day, so that they had extra time to work on their given routes. Estimated time of arrival at each pinpoint were carefully calculated and if any pinpoint was not located on ETA, a square search was to be made before proceeding to the next pinpoint. If necessary, to aid map reading, pilots would be permitted to climb to 500 feet shortly before reaching each pinpoint.

Gibson had listed the pilots of each group:

First group:	Himself,	Hopgood,	Martin,
	Young,	Astell,	Maltby,
	Maudsley,	Knight,	Shannon.
Second group:	McCarthy,	Byers,	Barlow,
	Rice,	Munro.	
Third group:	Townsend,	Anderson,	Brown,
	Ottley,	Burpee.	

The crews were:

Pilot: W/C G.P. Gibson DSO, DFC
F/Eng: Sgt J. Pulford
Nav: P/O T.H. Taerum
W/Op: F/L R.E.G. Hutchinson DFC
B/Aim: P/O F.M. Spafford DFM
F/Gnr: F/Sgt G.A. Deering
R/Gnr: F/L R.A.D. Trevor-Roper DFM

F/L J.V. Hopgood DFC
Sgt C. Brennan
F/O K. Earnshaw
Sgt J.W. Minchin
P/O J.W. Fraser
P/O G.H.F.G. Gregory DFM
F/O A.F. Burcher DFM

Pilot: F/L H.B. Martin DFC
F/Eng: P/O I. Whittaker
Nav: F/L J.F. Leggo DFC
W/Op: F/O L. Chambers
B/Aim: F/L R.C. Hay DFC

S/L H.M. Young DFC
Sgt D.T. Horsfall
Sgt C.W. Roberts
Sgt L.W. Nichols
F/O V.S. MacCausland

F/Gnr:	P/O B.T. Foxlee DFM		Sgt G.A. Yeo
R/Gnr:	F/Sgt T.D. Simpson		Sgt W. Ibbotson
Pilot:	F/L W. Astell DFC		F/L D.J. Maltby DFC
F/Eng:	Sgt J. Kinnear		Sgt W. Hatton
Nav:	P/O F.A. Wile		Sgt V. Nicholson
W/Op:	Sgt A. Garshowitz		Sgt A.J. Stone
B/Aim:	F/O D. Hopkinson		P/O J. Fort
F/Gnr:	Sgt F.A. Garbas		F/Sgt V. Hill
R/Gnr:	Sgt R. Bolitho		Sgt H.T. Simmonds
Pilot:	S/L H.E. Maudsley DFC		F/L L.G. Knight
F/Eng:	Sgt J. Marriott DFM		Sgt R.E. Grayston
Nav:	F/O R.A. Urquhart (DFC)		F/O H.S. Hobday
W/Op:	Sgt A.P. Cottam		Sgt R.G.T. Kellow DFM
B/Aim:	P/O M.J.D. Fuller		F/O E.C. Johnson
F/Gnr:	W.J. Tytherleigh (DFC)		Sgt F.E. Sutherland
R/Gnr:	Sgt N.R. Burrows		Sgt H.E. O'Brien
Pilot:	F/L D.J. Shannon DFC		F/L J.C. McCarthy DFC
F/Eng:	Sgt R.J. Henderson		Sgt W. Radcliffe
Nav:	P/O D.R. Walker DFC		F/Sgt D.A. McLean
W/Op:	F/O C.B. Goodale DFC		Sgt L. Eaton
B/Aim:	F/Sgt L.J. Sumpter		Sgt G.L. Johnson
F/Gnr:	Sgt B. Jagger		Sgt R. Batson
R/Gnr:	P/O J. Buckley		F/O D. Rodger
Pilot:	Sgt V.W. Byers		F/L R.N.G. Barlow DFC
F/Eng:	Sgt A.J. Taylor		Sgt S.L. Whillis
Nav:	P/O J.H. Warner		F/O P.S. Burgess
W/Op:	Sgt J. Wilkinson		F/O C.R. Williams (DFC)
B/Aim:	Sgt A.N. Whitaker		Sgt A. Gillespie (DFM)
F/Gnr:	Sgt C. McA. Jarvie		F/O H.S. Glinz
R/Gnr:	Sgt J. McDowell		Sgt J.R.G. Liddell
Pilot:	P/O G. Rice		F/L J.L. Munro DFC
F/Eng:	Sgt E.C. Smith		Sgt F.E. Appleby
Nav:	F/O R. MacFarlane		F/O F.G. Rumbles DFC
W/Op:	Sgt C.B. Gowrie		Sgt P.E. Pigeon
B/Aim:	F/Sgt J.W. Thrasher		Sgt J.H. Clay
F/Gnr:	Sgt T.W. Maynard		Sgt W. Howarth
R/Gnr:	Sgt S. Burns		F/Sgt H.A. Weeks
Pilot:	F/Sgt W.C. Townsend DFM		F/Sgt C.T. Anderson
F/Eng:	Sgt D.J.D. Powell		Sgt R.C. Patterson

Nav: P/O C.L. Howard
W/Op: F/Sgt G.A. Chalmers
B/Aim: Sgt C.E. Franklin DFM
F/Gnr: Sgt D.E. Webb
R/Gnr: Sgt R. Wilkinson

Sgt J.P. Nugent
Sgt W.D. Bickle
Sgt G.J. Green
Sgt E. Ewan
Sgt A.W. Buck

Pilot: F/Sgt K.W. Brown
F/Eng: Sgt H.B. Feneron
Nav: Sgt D.P. Heal
W/Op: Sgt H.J. Hewstone
B/Aim: Sgt S. Oancia
F/Gnr: Sgt D. Allatson
R/Gnr: F/Sgt G.S. McDonald

P/O W. Ottley (DFC)
Sgt R. Marsden
F/O J.K. Barrett (DFC)
Sgt J. Guterman (DFM)
F/Sgt T.B. Johnston
F/Sgt F. Tees
Sgt H.J. Strange

Pilot: P/O L.J. Burpee DFM
F/Eng: Sgt G. Pegler
Nav: Sgt T. Jaye
W/Op: P/O L.G. Weller
B/Aim: Sgt J.L. Arthur
F/Gnr: Sgt W.C.A. Long
R/Gnr: F/Sgt J.G. Brady

After he had had his say, Guy Gibson handed over to Barnes Wallis whose first words to the assembled airmen were: 'I feel very humble and honoured tonight, to you, who tonight will be going out and showing the world what the weapon will do.' Jim Clay of Munro's crew said at the time that it was strange that this quietly spoken, white-haired man, should have anything to do with bombs and destruction.

Wallis went on to patiently explain to the crews how the dams had been constructed and how important they were to German industry. He concluded by saying: 'You gentlemen are really carrying out the third of three experiments. We have tried it out on model dams, also a dam one-fifth in the size of the Möhne dam. I cannot guarantee it will come off, but I hope it will.' After the briefing, Wallis said to the people on the platform, 'They must have thought it was Father Christmas talking to them.'

As he finished his talk to the crews, the eight wheel trucks were rolling across the airfield, loaded with the huge cylindrical bombs, covered with a tarpaulin and still warm from the four tons of special high explosives put in at Woolwich Arsenal.

It was now the turn of the Chief Signals Officer who later had one final chat with all the wireless operators in batches according to the group they were flying with. He spoke about the radio signals and the use of both R/T and W/T on this operation.

The routes out and back were covered. The distance to the dams was 400 miles from Scampton. Great store was placed on avoiding the known anti-aircraft gun positions and the other likely obstacles. Wing Commander Percy Pickard DSO, DFC, one of Bomber Command's best known pilots, had helped Gibson plan the twisting course through the coastal defences, trying not to come any closer than a mile to any defensive positions. Hopgood had seen the plan earlier, noting that one position had been missed. He had come across it when he was on a raid while flying near the German town of Huls, where a large rubber factory was heavily defended. He was able to re-route that part of the course.

Reconnaissance of the dams had been taking place for many weeks, checking the level of the water in particular. On the 16th a reconnaissance aeroplane brought back photographs showing the level as being four feet from the top. It also showed that only the Möhne was defended; anti-torpedo nets in front of the dam, spread with a double line boom, and three light 3-inch anti-aircraft gun positions to the north of the dam, with possibly three or four light guns in the dam towers.

The met. officer predicted good weather over the target and that there would be a full moon. This would rise at 5 pm on the 16th, and set at 4.54 am on the morning of the 17th.

Jim Clay of Munro's crew recalls that there was none of the usual ribbing by 57 Squadron. They obviously sensed that a big job was on. After all the training 617 had undertaken, everyone was in high spirits and keen to get on with it. When the briefing was over, Dinghy Young went over to Wally Dunn and said how nice it was to hear one of his briefings again, having heard him on many occasions at Driffield back in 1940 while with 103 Squadron.

Out on the airfield the aircraft were almost ready. The weight set of the Lancasters, with fuel, the bomb and its crew was broken down as follows:

(*Left*) Flight Sergeant 'Chiefy' Powell (*Right*) Wing Commander Dunn taking the signal Nigger

Nigger's grave, Scampton

Nigger
The grave of a Black Labrador
Dog Mascot of 617 Squadron
owned by Wing Commander
Guy Gibson, V.C., D.S.O., D.F.C.
Nigger was killed by a car
on the 16th May 1943. Buried
at midnight as his owner
was leading his Squadron
on the attack against the
Mohne and Eder Dams

Lancaster and crew	–	40,000 lbs
Petrol (1740 gallons)	–	12,550 lbs
Oil (150 gallons)	–	1,350 lbs
Bomb	–	9,100 lbs
	total:	63,000 lbs

The bomb had originally weighed 11,600 lbs but had been reduced by the removal of the wood covering.

The only incident to mar the proceedings had been the death of Gibson's dog Nigger the night before. Usually to be found close at Gibson's heel, Nigger who in any event had virtual freedom of the whole aerodrome, was run over at the main gate at Scampton. It was a very busy road and it was believed that he was run over by a taxi. Being night and with Nigger very dark in colour it is likely the driver failed to see him until it was too late. Chiefy Powell told Gibson's batman Crosby not to tell the 'boss' until after the raid, but Crosby let it slip out. Chiefy went down to the guardroom when Nigger's body had been placed. He found him in one of the cells. He had been killed instantly.

  Gibson was naturally upset. Nigger had been with him a long time and they had been through much together. He asked Chiefy Powell to bury him, asking him to perform this task sometime after midnight. 'About this time,' said Gibson quietly, 'we should be over the target.'

CHAPTER SIX

'Enemy coast ahead'

The 16th May 1943 had been a fine day, and a busy one. The usual activity of a busy RAF bomber station was felt and heard. To get a squadron into the air for any operation, let alone this special one, relied on a great many people, many of whom were rarely seen or heard. It was a mammoth team effort. Fitters, electricians, riggers, armourers, parachute workers, flying controllers, intelligence staff, the met. people, equipment, clerks, security police, cooks, mess staff, transport – all had a job to do, each became a vital cog in the vital wheel of war. At Scampton on that day, nineteen Lancasters had to be made ready for one of the most unusual but spectacular raids of that war.

The staff of the Officers' Mess at Scampton knew all the tell-tale signs of a big day; two eggs and bacon for all flying personnel, which was served up by Leading Aircraftwoman Edna Broxholme and fellow waitress Leading Aircraftwoman Norma Hubbard. The crews themselves had spent the afternoon on last minute adjustments, checking guns, instruments, equipment, etc, and all the last little things that could make or break a raid of this magnitude. For many there was a few quiet moments in which to write a letter or two – to be left with the adjutant or the padre – just in case. Perhaps too a visit to the Station Chapel.

For Gibson's flight engineer (tonight would be his eleventh trip to Germany) John Pulford, there was a special compassionate leave off the Station. His father had died and he attended the funeral in Hull. He was escorted by two RAF policemen and was not allowed to mix with the people there just in case he let something slip to do with the operation that night.

At about 8.30 pm the crews were out at their aircraft awaiting the order to climb aboard. In the case of the crew of M-Mother – John Hopgood's boys, the time was spent in a friendly game of cricket, with the exception of the tail gunner and bomb aimer,

Tony Burcher from Australia, and Jim Fraser, from Canada. They were happy just to sit by the Lancaster and reflect on what lay ahead. Burcher had already completed a tour with 106 Squadron, and Fraser had flown ops. with 50 Squadron. As the events of the night were to turn out, it was strange that it was these two who had not joined in the cricket game.

When finally the order came to load up, the bomb aimer went in first. He entered the aircraft by the starboard rear door, up a small rung ladder, turned right to make his way up towards the nose. To do this he had to climb over two supports in the middle of the aeroplane, the first being two feet high, the second, the famous main spar, being about five feet high. In full flying kit, this was no mean feat. He went past the wireless operator's and navigator's tables dropping a step as he squeezed by the engineer, into the front of the bomber's nose compartment. Here he could lay prone on his padded rest under his chest. On his left was the bombing panel with its many switches and controls. Above him had wriggled the front gunner into the forward turret. He was seated behind two Browning .303 machine-guns, with his feet in the two special stirrups to keep them out of the way of the bomb aimer. The next man was the pilot who went past the wireless operator's and navigator's tables and turned left and into his padded seat that would be his home for some eight or nine hours. It was a little awkward to get into this seat but once settled in it was quite comfortable, and, as one pilot put it, a little like being ' ... on top of Mount Everest'.

The next man up was the flight engineer who took the same route and sat on a folding seat to the right and next to the pilot's seat. On take-off and landing it was his job to assist the pilot with landing gear, flaps and assist with throttle control. During the flight he would keep a watchful eye on the many dials and gauges on a panel alongside his seat, which dealt with engines, fuel, oil pressure, etc. All the while his experienced ears would be listening to the pitch of the four Merlin engines.

Up next came the navigator to take his seat behind the pilot at a table about three feet by four. The next man, the wireless operator, sat on a seat behind the navigator, his radio equipment separating them. He had the morse key at his right hand and to his left there was a small window which could be covered with a blind or curtain.

Through this window he had an excellent view of the two port engines.

Finally the tail gunner. He turned left inside the fuselage and made his lonely way into the rear turret with its four .303 machine-guns, first passing the Elsan portable toilet. On through two folding doors which were designed to keep out the draught but, more importantly, fire. He then climbed over the tail-plane support which was about two feet high and on through an opening of about three to four feet, into the turret. To say the least it was very cramped and despite the doors, the heavy flying clothing, sometimes electrical suits, etc, – usually bloody cold! It was made more so by the experienced gunners who would have had the glass panel immediately in front of them removed in order to be able to look straight out into the night sky. To see the enemy first was to survive – possibly!

One of the first aircraft ready for take-off was Joe McCarthy's. He then received a red Very light to start engines and proceed, but on trying to start up, he found one engine to be malfunctioning. He and his crew quickly left 'Queenie' and made for the spare aircraft, T-Tommy. Dave Rodger climbed into the rear turret only to find that the glass panel of this aircraft was still in place. He yelled to one of the ground crew who helped him to smash it out. This done he loaded the guns but he felt uncomfortable in this strange turret. The only consolation, however, was that on this trip, being flown at low level, he would not have to worry too much about the cold. Dave Rodger had come a long way to be with 617 on this night. Initially turned down by a Canadian recruiting officer, he had finally made it, including a necessary nose operation required by the medics, which he paid for himself. After training he eventually arrived in England in March 1942 and then on to 97 Squadron in October. In a crash on 31st October he had a kneecap smashed but he rejoined 97 after a spell in hospital, going into McCarthy's crew.

Joe McCarthy settled down into his strange cockpit, only to discover that the compass card was missing. In the panic to sort out a new one his parachute rip-cord was pulled, spilling his seat-pack 'chute. However, Chiefy Powell was able to get him another one on time, but the delay put them back. Instead of being first off, he eventually got off the ground at ten o'clock – thirty minutes late!

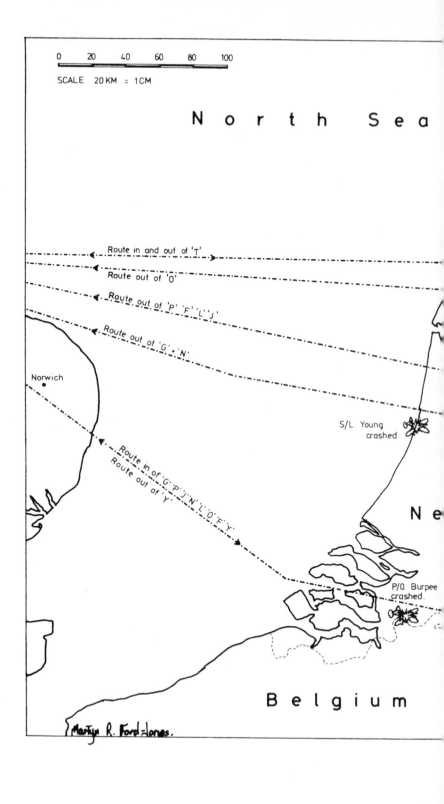

0   20   40   60   80   100

SCALE   20 KM = 1CM

N o r t h   S e a

Route in and out of 'T'

Route out of 'O'

Route out of 'P' 'F' 'L' 'J'

Route out of 'G' + 'N'

Norwich

S/L. Young crashed

Route in of 'G''P''J''N''L''O''F''Y'
Route out of 'Y'

N e

P/O. Burpee crashed.

B e l g i u m

Martyn R. Ford-Jones.

V.W. Byers crashed

G e r m a n y

• Zwolle.

l a n d s

S/L. Maudsley crashed.

• Munster

F/L Barlow crashed

F/L. Astell. crashed

Dorsten •

P/O. Ottley crashed

Hamm •

Gladbeck •

Dortmund •

F/L Hopgood crashed.

Moehne (X)

Hagen •

Sorpe (Z)

Remscheid. •

Schelme (E)

Eder (Y)

The second group, who were flying to the Sorpe, were away first. E-Easy, flown by Barlow left the ground at 9.28 pm. Behind him came Les Munro in W-Willie at 9.29. Third away was Vernon Byers in K-King at 9.30, and last was Geoff Rice in H-Harry one minute later. As events were to prove, none of these aircraft was to reach the target, and only the delayed McCarthy, the group leader, would make it.

Gibson's lead group began to take off next, heading for the Möhne and Eder dams. The first section of three took off together at 9.39 pm. In his log-book 'Terry' Taerum, Gibson's Canadian navigator, recorded the time as 9.40. Gibson had fired a green flare before take-off, a signal all was well and clear. With him went Hopgood's 'M' and Martin's P-Popsie (previously Peter). The next section roared down the runway at 9.47: A-Apple with Dinghy Young, J-Johnny flown by David Maltby and L-London (but known as 'L for Leather' to its crew) with Dave Shannon. The last of the nine took off at 9.59: Maudsley in Z-Zebra, Bill Astell in B-Baker and lastly Les Knight flying N-Nuts.

Corporal John Bryden, who had joined 617 from 57, watched the aircraft depart and fly low over Scampton, then formate. Also watching were Leading Aircraftmen Nick Carr and F. Payne who had serviced Gibson's G-George, and Young's J-Johnny. They watched the Lancasters circle once and then disappear in the distance. Another watcher was Leading Aircraftman Keith Stretch. He should have been on duty but had organised a stand-in for the evening. He watched them go, then he left the base. It was well past midnight before he arrived back, climbing through a hedge at the back of the airfield. He was met by his stand-in, Ray Fisher, who told him very excitedly, 'They have gone and I think it's the real thing!'

The thoughts on most of the ground crew's minds this night was, where have they gone and what was their target? One other who had thoughts of his own was Sergeant Jim Heveron in the Orderly Room. One of his tasks was to accept the wallets and the next-of-kin letters from the aircrew for safe keeping. He locked them all in his office safe.

Also in the night sky were the diversionary and other raids. Three

DH Mosquitos of 2 Group had been assigned an attack on Berlin, two more for Kiel, two for Cologne and two others to bomb Münster. Two Stirlings of 15 Squadron, four from 75 Squadron, two from 49 Squadron and six from 218, together with two Lancasters from 115 Squadron, were laying mines off the Frisian Islands – an operation known as 'Gardening'. Also laying mines were eighteen Wellingtons from 196, 431 and 466 Squadrons, operating off Brest, Lorient and St Nazaire. No 290 OTU, part of 92 Group, had four Wellingtons dropping leaflets over Orleans. Two Halifax aircraft of 138 Squadron on Special Operations (agent and resistance work) were operating over France, while another Halifax of 161 Squadron was on similar work over Denmark. It was not usual for Bomber Command to launch large-scale attacks during full moon periods for obvious reasons, so the Mosquito raids were usually planned for their nuisance value. Mine-laying all along the hostile coast would help to cover the low flying Lancasters of 617. If seen racing low over the coast, observers might think they were lost or a little off course, but also on mine-laying jobs.

As 617 Squadron flew out, drifts were used on the first leg across the North Sea. Flame floats were used, also GEE (*), by the navigator on part of the sea crossing. Many used GEE as a check over the continent as well. Fixes were obtained up to 0651 E, and one navigator used the set for homing on the return route.

It did not take long to cross the sea. Soon the pilot, bomb aimers or front gunners spotted the hostile shore. Not for the first time came the words, 'Enemy coast ahead, Skip.' A simple phrase, but one packed with thoughts, feelings and emotions. From now on people would be trying to kill them.

The second group, timed to cross the coast at the same time as the lead group but at a different spot, were the first to run into trouble. Geoff Rice saw gunfire and saw a Lancaster hit by anti-

(*) GEE – The aircraft's set received impulses sent out from a master station in England and measured the time interval of the triple signal reception. The GEE chart enabled the navigator to convert the time into distance and to fix his position. Navigation was quick and accurate with GEE to a range of 350 to 400 miles.

aircraft bursts as it roared over Vlieland on the Dutch coast at 300 feet. He watched it go down and crash at Texel. The time was 10.57 pm.

In fact the Lancaster crashed in the Waddensea a few miles east of Texel, killing all the crew. It was Sergeant Byers. He would never fly his mother over the beautiful English countryside. His bomb exploded some weeks later.

Geoff Rice was himself in trouble as he reached the Zyder Zee. Undoubtedly shaken by the sight of one of his companions going down so soon and so quickly he was hugging the surface of the water. It was very dark, unsuspecting and flat. He dropped down to 60 feet, then lower. Suddenly he hit the water. Instantly he pulled up but the damage was done. The belly of the Lancaster had been torn out and with it had gone the bomb. He shipped a good deal of water to the extent that his rear gunner, Sergeant Burns, was up to his knees in water. Burns at one time saw the water nearly envelop his turret.

The whole incident was one of sheer bad luck. At that height, in night-time conditions over a hostile countryside, it could have happened to anyone. He returned to base safely but obviously very dejected.[1]

Meanwhile, Gibson and his leading section had also reached the coast. The plan han been for the group to fly between two Dutch islands at Haltern off the coast of Holland, although both were known to have heavy anti-aircraft defences. Predictably Gibson came under fire from light flak guns and all three aeroplanes were picked up by searchlights. Gibson put his aircraft into a severe turn out of the beam before he continued towards the target, racing low and fast over the flat Dutch countryside.

Hopgood had also been caught in the searchlights, taking equal evasive action. Tony Burcher, just moments before, had been thinking of his bride-to-be Joan, whom he was to marry on 12th June. Just for those moments his mind had been miles away but he

---

[1] The bomb which was torn from Rice's aircraft was said to have been found in April 1973 at Oudeschild, but on checking with the Royal Netherlands Navy, they say the bomb brought ashore by fishermen was only 4,000 pounds. They defused it and later destroyed it at sea. I very much doubt if this was the bomb carried by Rice. Geoff Rice died in November 1981.

was quickly returned to reality as the Lancaster bucked up and turned, and he heard Pilot Officer George Gregory DFM, from Glasgow, in the front turret shout, 'Bloody Hell!' Tony said of that moment that his stomach did not belong to him and thinking any moment the Lancaster would hit the ground, but Hopgood got it under control and steadily climbed it on full throttle. For a brief second Burcher saw a looping arc of high tension cable just above his line of vision. Facing aft he saw the cable seemingly drop away behind the aircraft as Hopgood gained height. 'Right under the bloody things!' exclaimed Gregory.

'Sorry about that,' said John Hopgood.

Burcher muttered to himself, 'Good on you sport, good on you stone, old sport.' In his pocket he carried a stone for good luck. It had been given to him back in January by a little boy who had trotted up to him in a Lincoln Street and asked, 'Are you a flying man, mister!' 'Yes, son,' said Tony Burcher. The boy then said, 'The Germans killed my mum and dad in an air raid.' 'Well isn't that a damned thing,' replied Tony sadly. The boy then pushed a stone into Tony's hand and said, 'Next time you are over Germany, mister, throw it at the Germans for me. I hope it kills some bastard.' Tony never did throw it but kept it as a lucky charm through many trips.

Hopgood ordered Tony to keep his eyes peeled. His four guns were loaded with tracer as all the guns of the squadron were, instead of the usual one tracer shell to five bullets. This had caused quite a stir at the gun butts at Scampton and Tony had more stoppages (a side effect of tracer) than usual but on this night he was to use his guns much more than on any other operation. He opened up on the searchlights, but was unable to use his reflector sight as the bulb blew.

Suddenly the Lancaster was raked from nose to tail by ground fire; Tony was hit in the groin and stomach by shell splinters. The smell of cordite was in the air; 'When you feel the flak it's near,' recalls Burcher, 'when you smell it, it's too damned near!' A searchlight blazed full into his face and he fired a long burst at it, whereupon, the light went out.

'Got the bastard,' he yelled. 'Got him.' But at that moment a shell burst alongside his turret as the aircraft was swinging wildly.

He heard the flight engineer say, 'The port outer's gone, Skip, oil coming out and burning like Hell!' 'I'm feathering,' said Hopgood.

Burcher tried to rotate his turret but nothing moved as the hydraulics that operated it were worked by power from the port outer engine. His turret was out of action although he could hand-crank it. He felt saliva in his mouth, but then discovered it was blood.

Hopgood regained control and called each member of the crew on the radio. Sergeant John Minchin, the wireless operator, reported being hit. 'I cannot move my leg, Skip.'

Burcher reported that he too had been hit. There was no reply from Gregory in the front turret. Burcher heard nothing more from him and has always assumed that he had been either killed or seriously wounded.

Hopgood had been hit himself – in the head – and blood was pouring out of the wound, but he yelled to the engineer, 'Carry on and don't worry.' Burcher then heard him ask the engineer to hold a handkerchief to his head. He then heard the engineer exclaim, 'Christ, look at the blood.'

'I'm OK,' said Hopgood.

Flight Lieutenant Micky Martin in Popsie was still flying on track. The flak had died away and he flew very low but always giving enough room to turn and miss any obstacles that loomed ahead out of the darkness. He flew with minimum light in the cockpit to preserve his 'night vision'. His windscreen was smeared with salt spray, grime and insect carcases that had burst upon it. This sometimes confused a distant obstacle with a 'bug' but by moving one's head from side to side, bugs on the windscreen appeared to move while distant objects stood still. The Lancasters flew on, Hopgood still with them.

Back at Scampton Flight Sergeant Powell and Corporal John Bryden went down to the guardroom and collected Nigger's body. It was midnight and, as promised, they were going to bury him. This they did in front of No 2 Hangar, and within sight of Gibson's office window which was on the first floor, above the operations room. The grave was marked with a simple wooden cross made by one of the 'chippy' riggers, (woodworker). The grave can still be seen today, in the exact spot where Powell and Bryden buried him, but is now marked with an inscription. It was rumoured that the grave was later moved, but Chiefy Powell has visited the spot several times and has assured the author that it is in the same location.

Shortly after midnight, Gibson came under fire again. At 00.11 am he radioed back to Group HQ, 'Have been fired upon at position 5148N, 0712E.' Moments later some of the other Lancasters came under fire. Flying in formation with Maudsley, Bill Astell became uncertain of his whereabouts and reaching a canal crossing at the right place turned south down the canal as though to search for a pinpoint. Flying Officer Knight and his navigator, 'Hobby' Hobday, both saw Astell's Lancaster come under fire from the ground, his gunners returning fire, but then an explosion occurred and seven men died. The bomber crashed at Estate Achling Aarbek, ten kilometres over the Dutch/German border, just north of Dorsten, at fifteen minutes past midnight. Avoiding the ground fire, Astell had hit a high voltage cable, breaking the point of the pylon which was 30 metres high; the Lancaster hit the ground 500 yards away and blew up. The gunfire had come from Dorsten aerodrome, and although the explosion must have been seen, a crash report was not sent to the police until 3.59 am. Gibson's group was now down to eight aircraft, including the damaged aeroplane flown by the gallant Hopgood. The target was just minutes away.

With the lead group nearing the Möhne, Les Munro and Robert Barlow were still on route for the Sorpe, the only survivors of the second group, although McCarthy was behind them somewhere desperately trying to catch up. The overland route to the target consisted of short dog-legs, of perhaps fifteen miles, planned to avoid flak positions. GEE was working well, quite a long way into enemy territory.

Munro had flown to the southern end of the island of Vlieland where he changed course to the south-east, across the Zyder Zee. No sooner had they done this than a flak-ship opened up as their

Lancaster flew by, clearly silhouetted against the night sky. A hole was torn amid-ships, the intercom put out of action and the W/T rendered useless. Munro continued on for some time, trying to decide what to do. He decided to ask the opinion of the crew and had his engineer, Frank Appleby, pass a note round, saying 'Intercom useless, should we go on after all the training and hard work?' Jim Clay, the bomb aimer, wrote back, 'We will be a menace to the rest of the force with no intercom and no way of communicating with the rest of the aircraft.' This, of course, was vital if the operation was to succeed, and the safety of the other crews, and their own, had to be considered. Bill Howarth, in his front turret, felt the aircraft turn to the left and asked Jim Clay if they were returning to base? Jim nodded – they were going home.

When they reached Scampton they were met by Chiefy Powell and Jim Heveron. Both men saw what they thought was the bomb without its arm clamps on each side, and made a hasty retreat. Capable Caple meantime, rode up on his bike and told them it was only the fuel pipes hanging down and not the bomb arms. Munro had jettisoned fuel on the way back to lessen the load. They landed at 00.36 am – while 400 miles away, Gibson was making his bomb run.

Guy Gibson reached the Möhne dam at 00.15 am and ordered the force under his command to stand-by, and John Pulford pressed the switch to start their bomb revolving. This done, Gibson radioed, 'Hello all Cooler aircraft, I am going in to attack.' Terry Taerum, responsible for the height, switched on the two spotlights as they made their approach, at 00.25 am. Looking down he could clearly see them on the water but were still some distance apart. The self-destructing bomb fuse was activated on the run in by Spam Spafford in the nose, while Pulford monitored the speed at about 232 mph. Spafford then began to line up his special bomb-sight onto the dam's twin towers which they were fast approaching.

On the dam itself was Colonel Karl Burk, commander of the SS Flak Unit based there. One of his men, Karl Schutte, was on duty in the north tower and remembers:

'It was quiet. Apart from the regular steps of the sentry in the first tower above us, nothing could be heard. Then the telephone

(*Left*) Flying Officer Gillespie
(Barlow's crew)
(*Below*) Barlow's crashed aircraft

Barlow's bomb captured intact by the Germans

rang; Air Raid Warning! It took only seconds and the guns were ready to fire. We waited. We looked across the glittering lake where there was a reflection of the moon. Engine noises – their getting closer – low flying machine in direction X. It cannot yet be seen, the moonlight was blinding. The guns were ready, and the 2 cm guns barked away. The machine returned our fire and other machines fly in. The machines dropped flares and we were blinded as the first machine banked and flew low from the lake side over the wall. All three guns began firing as the first bomb fell short of he wall. Then a gigantic water spout rose into the air and waves swept over the wall. Now we knew the attack was meant for us although we had no time to admire this original drama!'

As soon as Gibson's Lancaster came within range of the flak guns, George Deering in the front turret opened up on the towers. The flak seemed accurate but it failed to hit the aircraft. The powerful retaining spring opened as the bomb release button was activated and the backwards spinning bomb fell away at 00.28 am, rotating at 500 rpm. Relieved of the weight, the Lanc lurched upwards, Gibson continuing the movement as he sped up and over the dam wall. The bomb bounced three times and appeared to slam into the parapet dead on target between the two towers. Hutchinson fired a red Very flare over the dam as a signal to the others that they had dropped. Passing over the dam, Gibson's rear-gunner, Trevor-Roper, began to fire on the gun towers.

Gibson hauled the Lancaster round in a tight circle. Looking down at the dam he saw the huge column of water rise into the air with water spilling over the wall, but as it fell back, he could see no sign of any breach. It was obvious that it was going to take more than one bomb to do the job.

The Lancaster force had also lost another aircraft. At precisely the moment Gibson's bomb went down, 00.28 am, Flight Lieutenant Bob Barlow of the second group – and apart from McCarthy the only survivor of the second group still en route, went straight into a high voltage cable. The machine crashed south-east of Emmerich at a place called Hedden in Rees-on-Rhine. All seven men died instantly.

CHAPTER SEVEN

The Attack

At 5 Group Headquarters Operation's Room at Grantham, Sir
Arthur Harris and Ralph Cochrane had been joined by Barnes
Wallis, who had come straight from Scampton. Wallis immediately
asked if there was any news. 'Apart from a flak warning from
Gibson, nothing at all,' answered Cochrane.

One wall was dominated by a blackboard listing the aircraft
taking part in the attack. On a dais alongside the opposite wall sat
the operations officer who was in telephone contact with the radar
room. Then from Gibson came a signal to Group, using the pre-
arranged code: 'Goner – 68A time 0037.'

The number '6' signified Special Weapon released and exploded
five yards from the dam. The '8' meant no apparent breach seen.
The time was the timing of the signal. This signal was received by
Wally Dunn, who said to Harris, 'Sir, there is a signal coming in.
It's from Gibson's aircraft; it says, Goner – bomb released and
exploded five yards from the dam, no apparent breach. That's all.'

Nothing was said. Everyone keeping his thoughts to himself.

The flak gunners on the dam were not the only ones to realise that
the dam was the target, another being the foreman of the power
station below the dam, Herr Clemen Kohler. He heard from a look-
out at 00.20 of the arrival of possible hostile aircraft. He suddenly
realised that this was a night of a full moon, a night when the Royal
Air Force did not usually venture over the Reich, and he also
realised that the water level in the lake was higher than it had ever
been before. He telephoned the United Electricity Company of
Westphalia's Office at Neheim, a little town just down the valley.
He told them he thought the RAF were attacking the dams, but he
wasn't believed, so he put down the phone, opened the door to the
outside and looked out. As he did so, Gibson's Lancaster flew
overhead with all guns firing, then came the explosion and water
began spilling over the dam wall high above him. Kohler began to

run and didn't stop until he reached the side of the valley several hundred yards away. He then dropped down beneath a tree halfway up the slope where he looked back down to see if the dam was still intact. With some relief he saw that it was.

Gibson waited for the water disturbance to subside before calling up the next aircraft – Hopgood's damaged M-Mother. An additional value of Gibson's first run was that he was able to gauge the barometer pressure over the water at the prescribed height. Also judging the strength of the defences which seemed to amount to some fifteen guns, situated not only in the towers but on the banks on either side of the dam. Gibson said to Hopgood: 'Take over M-Mother. Good Luck.'

Tony Burcher in the rear turret of Mother had watched Gibson's attack as they circled around waiting either a result or a turn at attack. All he could think of was, let's get it over and get back! Hopgood called back to him, 'Stand by, rear gunner, they are putting up a barrage ahead.' Gibson continued to circle, drawing off some of the fire from Hopgood. He saw Hoppy's spotlights go on over the water and immediately attract all the fire power the Germans could muster; they were not going to be outdone this time.

Karl Schutte with the guns on the dam continues:

'Target change, new inflight. It roared towards us like a beast as if it would ram the tower and us. One did not think of the danger, at last we could fire. I stood behind the gunner adjusting the gun height and making corrections, at the same time adjusting the side directions. We fired – whatever the gun would give. The shells whipped into the face of the attacker.'

Burcher heard the shout from navigator Ken Earnshaw to 'Go lower, still lower!' as he watched the two spotlights join up. He then heard, 'Bomb gone!' from Fraser. Just at that moment there was a terrific crash and Burcher saw flames streaming past his turret on the port side.

Sergeant Brennan, the engineer, shouted: 'We're on fire, port inner engine.'

'Press the extinguisher and feather Number Two engine,' commanded Hopgood.

This worked for a second or two but then the fire relit itself. With the port outer already dead and now the port inner blazing, Hopgood had very few options open to him at this low height. Instantly he gave the order to prepare to abandon the aircraft.

While this was going on the bomb was bouncing towards the dam, but it had been released just a fraction of a second too late. In his rear turret, Burcher tried to swing his turret around, using the dead-man's handle to slowly hand crank it to the fore and aft position. He recalls doing this in record time as his parachute was inside the fuselage behind him and he could not reach it until the turret was in the correct position. Managing this, he pressed the door release, scrambled out and grabbed his 'chute, clipping it to his chest. He then plugged into the intercom.

'How are you doing up front?'

Hopgood yelled back, 'Get out you bloody fool. If only I had another 300 feet – I can't get any more height.' All this took place in about 25 seconds after passing over the dam wall.

The flak gunners on the dam gave a shout, 'It's on fire', as they saw their guns hit home and the aircraft catch fire. An eye witness saw the bomber fly over the dam followed by a giant mushroom of foam in front of the wall. Seconds later the detonation reached them, and the pressure was so great the watchers were flung from their open door. Suddenly a flame came from the bomber and looking like a giant torch it flew over the town of Haar, disappeared, and was then heard to crash. The bomb, meantime, had bounced over the wall of the dam and exploded, completely destroying the power station.[1] The explosion knocked the flak gunners off their feet, and seconds later nothing was left of it.

'The firing was good,' continues Karl Schutte. 'The plane burned and I shouted to my gunner and then came a crack – dust

---

[1] During the 1950's Tony Burcher met a German engineer who told him that 'some darn fool had blown up the power house which caused great problems for years to come.' It was not until 1953 that this power house was rebuilt, not in the same position as the former one but on one side of the dam as opposed to directly behind it. When Tony Burcher related this to Barnes Wallis at the premier of the film *The Dambusters*, his reply was that he had tried to get the RAF to bomb it from normal bombing levels since 1941 but that they had refused, and so the concept of the bouncing bomb came to be born.

took our breath away. Strong pressure threw me to the ground and stones flew about our ears. The power station stood no longer. A quick glance at the burning plane; a strong explosion confirmed our success, but we had no time to look around for the hot, glowing gun barrel needed changing and a quick oiling. Apart from the directional gunner and munitions loader, everybody was replenishing the magazines since practically all the shells had been fired. Everyone did his best. Suddenly a report came from Tower Two – "Out of Action" – the impact had flung it into the plinth.'

Hopgood's Lancaster crashed three miles to the north-west, near the village of Ostonnen, which is five miles west of Soest. It burst into flames killing all the crew left on board. As Hopgood struggled to keep it in the air in order for his men to get out, Burcher saw the wireless operator, who had been wounded over the coast on the way in, dragging himself the length of the fuselage towards the rear hatch. John Minchin's face was white with pain, his leg had been nearly severed. He had sat with this terrible injury for the past hour. All Burcher could do to help was to clip on Minchin's parachute and push him out into the darkness. As he did so, Burcher pulled Minchin's D-ring release, but he did not see if the parachute opened, although witnesses did see two parachutes deploy.

Burcher then pulled his own parachute release while still in the aircraft. He knew it was not in the text books, but at this height he felt it was his only chance. Bundling it under his arm he plugged in the intercom for the last time.

'Rear gunner abandoning aircraft,' he yelled.

Hopgood yelled back, 'For Christ's sake get out of here!'

At that moment there was a terrific bang and a great rush of air. The flames had reached the main wing fuel tank. Burcher was blown out and smashed into the tailplane so violently that he broke his back. To this day Tony Burcher still has a hollow in his back and the break can be felt and still seen on X-ray. He landed with a terrific thud, which was only to be expected at such a low height. As he hit, the parachute billowed and took him back up again and it was this, a German Medical Officer said later, was what saved him. Landing again he lay stunned, hearing the other aircraft overhead and the ground vibrating beneath him. Dan Walker, Shannon's navigator, had watched Hopgood's aircraft go down and was amazed to learn that Tony Burcher had survived. The Lancaster, he said, seemed to speed itself along the ground in flames.

Gibson spoke to Martin, ordering him in for his attempt on the dam. As Martin began his run, Gibson flew his aircraft alongside Martin's to help divide the fire power coming from the guns. Martin was hit several times, having his starboard outer fuel tank and ailerons damaged – luckily the tank was empty at this time. The front gunner, Tony Foxlee, was returning fire with his two guns, while Tammy Simpson in the rear, sprayed them as they passed. Martin was later to praise the work of both gunners, as well as the rest of his men in the wonderful show they put up. He was also grateful to Gibson, and the way he helped draw off enemy gunfire.

Martin's bomb went down at 00.38 although there was still a good deal of smoke from the demolished power station still in the air. The number of bounces could not be seen but there was another huge water spout, the ripples spreading out, with a huge wave spilling over the dam wall.

Karl Schutte: 'Now there were only two small guns to fight the beasts. The engine noise came nearer again. Banking planes dropped more flares and also showed their directional lights to draw our fire. We restarted firing which was returned by the machines. While they still tried to draw our fire, another plane raced towards the wall. Target change, and again the shells whipped towards the attacker and several hits were scored, but what could our 2 cm guns do against an armoured flying fort, it was just scoring good luck hits. The plane was now also shooting at us. Just like a string of pearls, the luminous spur of the shells came towards the tower like large glow-worms.

'Then came a heavy explosion and a great water spout, again the lake quaked and waves engulfed the wall. We did not know whether the wall was still standing, we must only fire and fire again. Then came the fourth attack and the picture was repeated but we didn't know which machine to fire at first. This attack too was unsuccessful. Could we do it, would we fight them with our

(*Top left*) Flying Officer (
(Hopgood's crew)
(*Top right*) Tony Burcher
(Hopgood's crew)

Hopgood's crashed aircr

barking guns and foil the attack? We hoped no other guns would be put out of action, since the three guns of the 2nd Unit were down in Gunne and could only fire on machines veering away, since the attack was only coming from the lakeside.'

The fourth aircraft Gibson ordered in was Dinghy Young's. He came in low, dropped his bomb at 00.40, while his gunners exchanged fire with the men on the dam. Once again the bomb bounced against the wall correctly and the now familiar water spout erupted and a huge wave of water spilled over the wall. Young yelled to Gibson that he thought the dam had gone, but Gibson looking down, replied, 'I think it will take one more bomb.' Young sent his message back to base at 00.50, code 78A – weapon release and exploded in contact with the dam wall, no breach seen. Mick Martin, who had bombed earlier, did not send his message back until 00.53 – code 58A, spun weapon, released, exploded 50 yards from dam, no apparent breach.

J-Johnny with Dave Maltby was next. This time there was no flak fire from the dam. Karl Schutte continues:

'Some machines were firing at us from the valley end and we replied, but then the gun failed – the lock was stuck. We tried desperately to remove the cause and even tried sheer force, but without success. A premature shell had damaged the housing. It was hopeless. We stood high on the tower, in front of us the lake and behind the valley. Attacked from all sides we could not defend ourselves. We waited literally for the end.

'Then a fifth plane started its attack. Only the gun on the lower wall was still firing. The machine neared the wall at an incredible speed; they now had an easy game – I could almost touch it, yet I think even today I can see the outline of the pilot. With our gun silent we did that which we had drilled so often – defence with rifles!'

Maltby's bomb was again on target and the water flew into the air. When it cleared a hole could be seen in the dam and getting bigger by the second. In fact Maltby had seen a minor breach in the dam on his run-in so it appears that Young's bomb had started the ball rolling. Maltby's wireless operator called base at 00.55 am, code 78A – weapon released, exploded in contact with the wall – no apparent breach.

Suddenly the water was rushing down the valley and the air was full of spray. At this precise moment, Gibson was flying around and could not see the explosion owing to smoke, and spray settling on his windscreen. Time was now getting short. He was about to order Dave Shannon to make his run when he saw a large hole in the dam and water pouring out. 'It has rolled over!' he yelled, then ordered Shannon to break off and not to attack. Taking a closer look, he saw a breach of 150 yards in the dam wall.

The men at Grantham were tense. What was happening? Would it work? They had just had the message from Maltby saying 'no breach', and he was the fifth to attack! Little was said. Undoubtedly Wallis was making and remaking mental calculations, his instincts saying it must work but his ears hearing messages saying it was not.

At 0056 am Gibson sent a message with the pre-arranged code word 'Nigger'. Wing Commander Dunn began to take down the signal. He got to 'N-I-G ...' and bellowed out at the top of his voice, 'Nigger!' Harris dashed to Wallis and shook his hand, patting him on the back, and said, 'I knew it would work.' At 00.57 Group replied to Gibson on full transmission, asking him to repeat his signal. Gibson did so with one word.

'Correct.'

On the dam Karl Schutte watched in horror. 'Again the lake quaked and a gigantic wave came over the wall. The wall had been breached and relentlessly the water began to run into the valley. The planes banked away.'

Leutnant Freswinkle had also been on the dam, and had given orders to his gunners to fire sustained automatic fire through all the apertures in the tower and wall, but it had all been to no avail.

Gibson circled for three minutes and then called up the other Lancasters, ordering Maltby and Martin back to base. This they did, landing back at 3.11 and 3.19 am.

Tony Burcher was still lying stunned on the ground. He looked towards the dam to see a great column of water. It went up like a giant soda syphon, followed by a roaring sound. He thought, 'I'm going to be drowned.' He was, of course, over a mile away but in his shocked state this did not register.

The power station foreman, Clemen Kohler, was still sitting beneath the tree, and watched as the masonry bulged then burst between the two towers. The first rush of water plunged down to the bottom of the valley, striking the ground with a colossal crash. The remains of the power station vanished in a second and the tidal wave settled in one gigantic rush, pouring down the valley at twenty feet per second, taking everything with it. Further down the valley, car headlights changed colour as the water overtook them.

The Möhne dam had been smashed.

Back at Grantham, Harris was anxious to give the news to the Prime Minister who was at the time in Washington, visiting President Roosevelt. Harris went into the silent glass telephone cabinet in the corner of the ops room, picked up the 'secret' telephone and asked the WAAF telephone operator to get him the White House. She turned to Wing Commander Dunn who reassured her that it *was* the White House in America that Harris wanted. She in turn asked the Bomber Command switchboard to get the call for her. The Bomber Command telephonist was a little sleepy in response, not knowing there were operations on that night. At first she did not take the request seriously but was soon told in no mean terms to get the call. Now wide awake she got a line straight through, via the American HQ in London. The American soldier on the other end was quite content to spend the rest of the war speaking to the sweet English female voice on the other end of his phone, but finally he was put through to Portal who was with the PM, but not before Harris had got a trifle hot under the collar.

In Neheim, just down the valley, the first air-raid warning came at 12.30 am, the duty officer at the police station; Leutnant Dicke, himself hearing aero engines, went to the watch tower at the town hall to assist in assessing the situation. At 00.15 he first observed a very bright light, and then a muffled explosion was heard. Some ten minutes later the telephone rang at the station. It was the police station at Arnsberg, enquiring about the bombs. Dicke replied that the bombs had not fallen in the town area, but in the direction of the dam. Some while later the architect at Neheim, telephoned to say there was a rumour that the dam had been hit, resulting in flooding. The observer station at the dam should have telephoned by now, but had not.

Then came a call from the 'Special Service' to Dicke, confirming the worst; the dam had been hit and water was pouring down the valley. This report was logged at 00.50 am. Dicke telephoned the Mayor telling him the news, who said he would come right away and to have one of Dicke's men meet him. A reserve constable was sent, but when he arrived at the Mayor's house, water was already surrounding it. Dicke then sent officers to the most vulnerable parts of the town to alert the population. Then the water got into the cellar of the post office where it destroyed the telephone system sometime before 1 am. Then the water really hit. The flood wave at Neheim was 39 feet. It raced down onto the town at thirteen miles per hour.

CHAPTER EIGHT

On to the Eder

At Scampton the reserve force had taken off as planned. Pilot Officer Warner Ottley in C-Charlie left the ground at 00.09 am, Pilot Officer Lewis Burpee in S-Sugar at 00.11,[1] Flight Sergeant Ken Brown in F-Freddie at 00.12, Flight Sergeant Bill Townsend in O-Orange two minutes later, and finally Flight Sergeant Cyril Anderson in Y-York at 00.15. Ken Brown reached the enemy coast at 1.30 am, Bill Townsend at 1.31.

At 1.45, Group transmitted a flak warning to the third group.

At the time Gibson and Co were attacking the Möhne, Joe McCarthy, the only survivor of the second group, arrived at the Sorpe without having encountered any problems. However, he found the lake difficult to find owing to mist. His run-in was made across the length of the dam. Because of the construction of the dam, an earth dam, the intention was to crack the centre, watertight concrete core so as to start a leak which might in time destroy the dam or at least force the Germans to empty the reservoir in order to effect repairs.

McCarthy made his first attack at 00.46, with the moon on his starboard side, but he was not happy with his approach and hauled off to come round again. In all he made some ten runs before release, which shows the tremendous dedication and patience shown by McCarthy and his crew in placing their special bomb in exactly the right spot. A spout of water rose 1,000 feet into the air. The resulting damage caused some crumbling of about 15 to 20 feet

[1] In navigator Tom Jaye's log-book the time was recorded as 21.29, but this was filled in by someone else after the raid, for sadly Jaye did not return to make the entry himself.

of the wall, necessitating the Sorpe dam being drained to half its capacity. There was no opposition to these attacks, the dam having no anti-aircraft defences.

For some reason, McCarthy failed to radio back of his attack until 3 pm, when they were only twenty minutes away from Scampton. Their message was '79C – Goner, Special Weapon released, exploded in contact with the dam – small breach in dam.' McCarthy had followed the planned route out without difficulty but was unable to return by the planned route owing to failure to find pinpoints because of a faulty compass. He therefore flew back via the outward route, the pin-point of the Zyder Zee being easier to recognise. They landed at Scampton at 3.23 am.

Having left the Möhne, Gibson and the remaining Lancasters of the first group were now nearing the Eder. With Gibson was Dinghy Young, deputy leader, (now bomb-less), Dave Shannon, Les Knight and Henry Maudsley. At the Eder they found no anti-aircraft defences but because of fog Gibson was unable to actually make out the dam for some time. He flew low and round trying to pick it out and was on his fifth run before he finally saw the reservoir, then he spotted the dam a few minutes later.

With the others now overhead, he radioed Dave Shannon to make his run. Shannon was also having difficulty in seeing the dam and was not even sure exactly where he was. Gibson told Bob Hutchinson, his wireless operator, to fire a red flare and at once Shannon called to say he was on his way.

The dam was in a deep valley amongst wooded hills. At the upper end of the lake, high on a hill of about 1,000 feet, stands the Castle of Waldecke, the home of Count Waldecke, which is now a splendid hotel and restaurant.

At 1.32 am a telephone rang in the local air raid defence control office. Leutnant Saahr of the SS answered. He was told, 'This is the Warnzentalle, there are British aircraft circling over the Eder dam.' The authorities of the valley below the Möhne dam had refused to believe the call from Clemen Kohler an hour before, but Saahr did not ignore this call. He rang the nearest SS unit, the Third Company of the 603rd Regional Defence Battalion, at Hemforth. The duty Colonel confirmed there were three (sic) aircraft circling the dam, and Saahr said, 'I will call you back in a couple of minutes. If an attack starts before I do so, sound the alarm.' He then telephoned SS Colonel Burke (not to be confused with Colonel Karl Burk on the Möhne dam), the Commanding Officer of the SS Flak Training Regiment stationed nearby, warning him that a flood was probable within minutes. Burke quickly had a hundred men with lorries stand by. Almost at once Leutnant Saahr telephoned again to give him the alarming news that, 'The local battalion says the planes are releasing flares, so we have switched on searchlights.'

Flight Lieutenant Dave Shannon made his run at 1.39 am by flying over the hill with the castle on top, and then diving steeply to the required height of 60 feet over the water of the reservoir. The approach was not satisfactory for the bomb aimer, 32-year-old Flight Sergeant Len Sumpter, who was on his fourteenth trip to Germany. Shannon circled back and was about to try again when Gibson called Henry Maudsley to make his run, so Shannon hauled off. Maudsley came in and dropped his bomb, which appeared to leave the aircraft late. It was later thought that his Lancaster might have been hit on the way to the target area which could have caused the bomb release to have been damaged. Something was seen to be hanging underneath the aircraft by the light of the moon, but it was not identified for certain.

The bomb bounced but overshot the dam and struck the parapet. The bomb detonated instantaneously as the Lancaster flew over it. Maudsley spoke briefly on the R/T but it sounded very weak. In the explosion the aircraft had been mortally hit and it headed away in the direction of Emmerich. Maudsley struggled with his crippled Lancaster for the next forty minutes. His wireless operator got off a radio message at 1.57 am, stating, 'Goner, 28B, special weapon released, overshot dam, no apparent breach.' Shortly after 2.30 am the Lancaster was nearing Emmerich, now on fire.

The light anti-aircraft post at Emmerich had strict orders not to fire on hostile aircraft so as not to give away the town's location. However, the aircraft, approaching from the south was recognised as British, flying low and on fire. One gun opened up on it and then the others joined in. The bomber was hit and went down, crashing three to five kilometres south east of Emmerich at a place called Netterden. There were no survivors.

At 1.51 am, Gibson took time to radio Astell, trying again at 1.53. He failed to get a reply. Astell had been dead for an hour and a half.

It was when he was making the second call, that Lewis Burpee was shot down on his way to the Sorpe, flying near Gilze-Reijen Luftwaffe base in Holland. He was seen to be a little off course, too far north by perhaps a mile. The flight plan had set a course to avoid the base but flying low, he was caught by searchlights, when only 25 metres above the ground. Dazzled and under fire from light flak positions, he dropped lower, attempting to avoid the probing lights and gunfire, but in doing so he struck some trees and crashed onto the air base, hitting an MT section housing trucks belonging to both fire and flak personnel. A few seconds later the Lancaster and its bomb exploded with an ear splitting explosion. Windows and doors of the base were blown in and smashed from the blast which was so great that the HQ building of the resident night fighter unit, NJG/2, some 6-700 metres away, was completely blown over.

The aircraft burned for some time, with smoke columns and flames reaching high into the air. Exploding ammunition aided the conflagration. The airfield's main flak building and command post, also the vehicle park, were all heavily damaged. All this was witnessed by a Ju88 pilot named Scholl, stationed at Gilze. Although Burpee and his crew failed to reach their target, they at least put Gilze-Reijen's radar post out of action for some time when they crashed.

When Maudsley had flown off after his attack, Shannon came in for his third attempt. Len Sumpter was happy with the height, distance and speed, and so released the bomb. It bounced twice and sank at the dam wall, followed by an enormous spout of water a 1,000 feet high, about a minute later. A gap of nine feet was seen towards the east side of the dam. His wireless operator, Flying Officer Brian Goodale DFC, signalled back, 'Goner 79B, special weapon dropped, small breach in dam.' It was timed at 2.06 am.

The next Lancaster in was Les Knight's. He attacked at 1.52, with the moon on his starboard beam. They flew a dummy run to make certain of clearing both the steep hill and the trees on the top. He then came in and released his bomb, having to hop over another very steep hill during the run, and flying very near the trees on the top. Gibson had flown alongside Knight and saw the bomb bounce three times, hit the dam, and explode. A message was sent by Knight's wireless operator, Pilot Officer Bob Kellow DFM, at 2 o'clock – 'Goner 71B, large breach in dam.' They made a circuit from a few hundred feet and were able to see the start of the flood in the bright moonlight. A spout of water 800 feet high spurted right out from the dam. It was a large breach thirty feet below the top, the water through which caused a tidal wave about thirty feet high and was seen half a mile down the valley from the dam. The water was smashing buildings and bridges, while further down cars were overtaken and lost to sight. The front of the flood was like the face of a cliff moving along and was practically perpendicular.

Gibson had in fact signalled back at 1.54 the call-sign 'Dinghy' meaning the Eder had been breached. It was re-transmitted by Group at full power for the benefit of the other aircraft, at 1.55 am.

With the aircraft still circling above them, the workmen standing on the steps of the generator plant below the dam, felt a dull shock and felt the building shake, thinking it would fall down. They immediately ran to the main room but the lighting had failed. Masonry came through the roof and then water flooded in, but they managed to reach the stone steps leading to the bank before the dam wall broke.

Five minutes after the Eder was breached, the telephone in Colonel Burke's office rang. It was Leutnant Saahr. 'Herr Colonel,' he said, 'Arolson Post Office has just phoned through a report from the 603rd Battalion, that the dam has been destroyed. I have tried twice to contact them but all the lines are dead.'

The village nearest the dam needed no telephone call to tell them about the breach. A motor cyclist rode through the main street of Affolden screaming at the top of his voice to the people taking cover following the air raid warning. 'The dam has been hit, the water is coming, everybody out of the cellars, quickly.' The villagers hurriedly evacuated their shelters and cellars, and headed for high ground. As they did so, Affolden vanished in a flood of water, while the street suspension bridge at Hemelfort collapsed with an enormous rumble.

At 2.10 am, Group HQ signalled to Gibson, enquiring, 'How many aircraft of the first group are available for 'C'?' (Sorpe dam). He replied at 2.11 – 'None.' The force under Gibson, set course for home, a journey in which they would now have to keep their eyes peeled for fighters and an alerted defence.

Dinghy Young had left the area of the Eder a couple of minutes earlier than Gibson and made his way towards home. He flew via the Ijeesellake Hoom Alkmaar, in the direction of Bergen, where the well-known defences gap, between Egmon and Scheer, was situated. Gibson flew more to the south on his trip back.

Young reached, then flew past the southern part of the gap and was a little higher than perhaps he should have been, which enabled German gunners to get his range. His Lancaster was shot down at Ijmuiden, west of Amsterdam. The aircraft started to burn as it lost height over the coast, then plunged into the sea close to the North Sea beach. It was 2.58 am.

Les Knight sped home at full speed, passing the Möhne dam on his way. By now it was half empty and the valley to the west was flooded. They were flying just behind Gibson although they could not see each other. In crossing the Dutch coast, Knight attracted some AA fire, so he flew a little further north to avoid the defences, and nearly hit a pole sticking up from the top of a hill. Clearing the enemy coast, he flew out to sea, setting a new course for base where he landed at 4.20. Shannon was ahead of him, landing at 4.06, while Gibson touched down at 4.15 am.

Having heard from Gibson that all aircraft of his group had bombed, Group signalled to each of the third group, who were still on their way out, at 2.21. Flight Sergeant Bill Townsend acknowledged at 2.22, and they then received the message, 'Gilbert – attack the target as detailed.' Townsend replied at 2.26, 'Message received.'

One minute later Group called Cyril Anderson, who replied, 'Carry on with your message.' Group told them, 'Dinghy – target breached, divert to target Z.' Pilot Officer Ottley's route to the

Lister dam was the same both ways. He was called by Group at 2.30 and the call was acknowledged. Group then told them, 'Gilbert' – the signal to divert to the Lister. Nothing further was heard from this aircraft. Near Heessen, north of Hamm, apparently a little off course, Ottley's Lancaster was hit by light anti-aircraft fire and went down. At 2.35 his Lancaster was seen to crash by Flight Sergeant Ken Brown, flying at 150 feet. He saw the bomber blow up with a large explosion.

Ken Brown and his crew had themselves just received Group's call telling them to attack the Sorpe, but had had an eventful trip already. They had shot up three trains, killing five German soldiers and wounding eight others. His Lancaster had been fired upon by flak, being hit in the fuselage, but nothing vital had been damaged.

He attacked the Sorpe dam at 3.14 am, the valley being hidden in a swirling mist. He made a total of eight runs but to no avail, but then dropped some incendiaries on the banks of the lake which set fire to some trees on both sides of the dam. Brown then tried another run, but he was still not satisfied. Then on his final run he saw the dam from 500 yards and dropped his bomb. He signalled back at 3.23, 'Goner 78C, special weapon released, exploded in contact with the dam, no apparent breach.' This attack and drop was the same as Joe McCarthy's, across the dam and without bouncing the bomb.

The bomb fell about ten feet away from the dam wall, about two-thirds the way across. A crumbling of some 300 feet was seen which enlarged the area of damage made by McCarthy nearly two hours earlier.

On Brown's return to base, he flew via the Möhne dam and his engineer, Flight Sergeant Basil Feneron, saw two large breaches close together, each about a quarter the width of the area between the two towers. Water was still pouring through both gaps, showering well out before falling in two powerful jets, and the valley seemed to be well covered with water. Both Sergeants Heal and Onacia, navigator and bomb aimer respectively, reported seeing a third breach between the towers on the north east end of the dam. Brown and his crew flew home safely, landing at Scampton at 5.33.

Flight Sergeant Cyril Anderson's target was also the Sorpe. He was unable to find the lake near Dulmen, mist in the valley making recognition difficult. About five minutes before they reached Dulmen they were coned by searchlights, but were unable to fire on them owing to stoppages in the rear turret guns. They began to realise they would not be able to reach the target on time, so turned back at 3.10 am, bringing their weapon back, landing at 5.30, coming home in the morning light. Wireless Operator Sergeant William Bickle signalled back at 4.23, 'Returning to base – unsuccessful.'

Flight Sergeant Bill Townsend headed O-Orange towards the Ennepe dam, on the river Schwelme. By this time it was well on into the early morning and navigation was very difficult over the hills as mist was forming in valleys and with the angle of the full moon the effect was to make it look like a lake. At last the dam was spotted by the profile of the hill three-quarters of a mile away. Townsend made three runs before the bomb-aimer let go the bomb at 3.37, and encountered no defence. The bomb bounced once and an explosion occurred approximately thirty seconds after release. The expectation was that at least one other of the reserve group would turn up, but none did.

A high column of dirt and water came up from the explosion but there was no sign of any damage being inflicted. They signalled back at 4.11 'Goner 58E,' which Group acknowledged one minute later.

Their journey back was flown with the sun coming up on one side and the moon going down on the other. High tension wires lay in wait for Townsend as he kept the Lancaster right down on the deck at about 240 mph, but it seemed to fox the enemy defences. They too flew back past the Möhne, although at first they could not see it until they saw a sheet of water about seven miles long and four miles wide. Back at Scampton Townsend was not at first believed when he reported the size of the area under water, but his bomb aimer, Sergeant Charles Franklin, confirmed it also saying that he had seen the roofs of houses sticking up above the fast flowing water.

Reaching the enemy coast, Townsend flew out between Texel and Vlieland. The defences on this part of the Dutch coast had a

(*Left*) The site of Astell's crash. (*Right*) Pastor Berkenkopf

Ottley's crashed aircraft

quick flurry at them but the Lancaster was not hit. They landed at Scampton in broad daylight at 6.15 am.

As the last signal came into Group HQ, Harris decided to go to Scampton and meet the returning crews. When he arrived, Gibson, Shannon and Martin were down. The ground crews had been waiting all night for their return, going for supper and then returning to their flight huts. Hours later the sound of aircraft was heard. As the Lancasters started to circle the field, it was realised they were 617 aircraft. Leading Aircraftmen Law and Payne were waiting at dispersal for Maltby's J-Johnny. The crew of Johnny alighted, full of the raid. The ground boys put the Lancaster away and then they went back to their billets for a few hours sleep.

Leading Aircraftman Keith Stretch waited for Les Knight's N-Nuts, and when they landed, Knight came over and said, 'I hope I haven't treated your engines too badly. They behaved beautifully but we did call upon emergency power for a short while.' Apart from this the Lancaster seemed to be in good shape. Corporal Chapman remembers two aircraft making abominable landings on the airfield, and also that there wasn't one aircraft that did not require some work before being able to fly again. Of the eleven aircraft that returned, four were damaged, three by flak and one by machine-gun fire. Gibson's G-George, remarkably, had only three small holes in its tail despite the length of time it had been exposed to the defences over the Möhne. In addition, Rice's aircraft had been damaged when it hit the water.

All the switches in the Lancasters were put to neutral, chocks put in place and only then did the ground crews report to the flight office, where they were told to return at 8 am the next morning. For some of the ground crews, however, there was eight ominously empty dispersal pens. As the time went on and it became lighter, it was obvious they would not be coming home.

In the Officers' Mess, Edna Broxholme, who just a few hours earlier had helped to serve the flyers their eggs and bacon looked sadly on all the empty chairs in the mess after the crews returned.

CHAPTER NINE

The Deluge

As dawn broke on the morning of 17th May 1943, the aircraft of the Minister of Munitions, Doctor Albert Speer, was landing at Werl, some distance from the Möhne dam. He saw the ruined villages, towns and cities and was told that the flood waters had reached parts of Kassel, thirty-five miles away from the rapidly emptying Eder dam, the largest tank and aircraft engine manufacturing centre in Germany. Two hundred yards of railway embankment had been destroyed at Wabern, the station sidings at Kassel were silting up and the dock railways under water. Parts of Kassel's U-boat, tank and artillery manufacturers had also been affected. These were all targets on which the RAF had made at least ten bombing raids without damaging them as much as the water had done.

The village of Gunne had been virtually washed away, also most of the town of Neheim-Hüsten. Here Josef Greis had risked his life to rescue children from an air raid shelter and raised the alarm, saving a number of people from certain death. Another, Johannes Kessler, had heard the water in the distance and sent one of his comrades to an air raid shelter where about 100 people were taking cover, and they too were able to get away in time. He then ran to houses where people were sheltering in cellars to warn them, finally working with the fire brigade and air defence workers, helping in clearance work. Both men were recommended for some form of war medal by the Mayor on 31st May.

At Froenberg, the canal had been destroyed, the power station knocked out and the railway bridge swept away like matchwood. The main railway viaduct that joined Dortmund and Hagen, had been severed and some thirty-two miles of countryside had been flooded.

Speer flew on to the Sorpe dam and ordered the pilot to land in the open countryside near the dam, against which two of 617's bombs had struck. Being a massive earth clay construction it did not lend itself to breaching in the same way as the Möhne and Eder had. When Speer arrived at the dam he was met by the local civil engineering chiefs.

They told him that repair of the Möhne was not going to be easy, for the breach was 250 feet at the top of the dam, 130 feet at the base. Speer telephoned an immediate order for anti-aircraft protection at every important dam in Germany – 'By tonight!' he yelled into the phone. Within twenty-four hours he had some 7,000 workers at the Möhne, workers taken from the construction of Hitler's Atlantic wall project. While there, these workers had to be accommodated in tents near to the dam.

The flood waters had caused much damage, from both dams. The village of Hemelfonten was a wilderness of ruins. The Pastor was Joseph Berkenkopf who had been born on 4th July 1880, becoming a priest in August 1903. In May 1943, he had been the Pastor of Hemelfonten for twenty-seven years. In his last sermon on Sunday 16th May, he read as his text, 'A short while and you will see me no more, for I am going to the Father.' When the water hit the village of Niederense the Pastor, like all the other villagers, was in the cellar attached to the church. After the first wave of water, the roof and tower of the church could still be seen, but when the second wave had passed over, all was lost to sight. The last tolling of the church bell could be heard and then it too became a victim of the war.

Siegfried Reinhold, who is today a policeman like his father before him, was a member of the Hitler Youth in 1943, and had been ordered to the church with others, to help dig for the bodies. Many small and large stones had to be moved, but finally they found the cellar. After shifting the sand they located the Pastor's body. Carefully lifting him out, he was placed onto a stretcher and taken away for burial. His text had become a reality.

The church collection taken that Sunday was found in September 1945. This, along with wooden statues which survived the deluge, now reside in the local museum at Niederense. The site of the church can still be seen from the outline of its foundations. Where the chancel had once been is a simple, large, wooden cross.

Möhne Dam after the raid

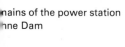ains of the power station
hne Dam

All that is left of a nearby farm is one cowshed; the site of the farm is now a play area and football pitch for children.

A farm worker in Niederense, Herr Kersten, gave the alarm as the water approached, calling, 'Save yourselves, the water is on its way.' Then he, with Frau Scheven, ran to the road which led eastward from the farm towards the woods. Between the buildings he came to the first waves of water, which were so high they covered the houses. Frau Scheven was dragged out of his hands along with a friend of hers, with her three-year-old child, who was on a visit. All three were carried away in the flood. Fraulein Muller, the owner of a chicken farm next door to the vicarage, saw the farmhouse and vicarage disappear first, the church and its tower standing for a further quarter of an hour then they too collapsed. All one could see was rubble, sand, mud and ruin as it got light. Drowned cattle, hens, pigs and occasionally human beings, could be seen hanging in the tops of trees. An old lady of eighty-eight, tied herself to the chimney of her houses and spent two days there before being rescued.

In a cemetery at Affolden, there are thirteen graves all with the date 17th May 1943. Three of the graves are for Gunter Neis aged ten. Matilda Laborez, aged three, and Konrad Buttcher aged seventy-four. Their graves, with the other ten, are of small marble stone, surrounded by many flowers. In this cemetery is a simple wooden cross in memory of the victims of the Eder dam flooding.

The valley of Affolden, Bergheum, Giflitz and Mehlen had been inundated. Homes were swept away while a side of the valley was eaten away, causing whole roads to disappear.

SS Colonel Burke, now in charge of the rescue troops, reported: 'The first impression of the damage is devastation.'

The work of the rescue squad became rougher. Thousands of cattle had to be dug out of the mud and buried. A state of emergency was declared in Westphalia at 4 am on the 17th. Shipping in the inland docks at Duisberg was ordered out into the comparative safety of the broad Rhine waters, south of the Admiral Scheer bridge. The Paul Baumer airfield on the south bank was evacuated of all stores, equipment and aircraft. Neheim, fifty miles below the Möhne dam, was engulfed in the early hours of the 17th, swamping the town's important coalfields and ironworks, forcing most of the 130,000 inhabitants to spend the night camping out on high ground above the valley. Earlier, the raging Ruhr had swept over Dortmund, five miles north of its normal course, flooding all the city's air raid shelters, causing incalculable havoc and driving more than 40% of the inhabitants into the cheerless night. Every town and village between the Möhne and Duisberg felt the fury of the flood.

The loss of life at Neheim alone was 51 men, 66 women and 30 children. Forty houses were destroyed, nine severely damaged, twenty badly damaged and a further 177 slightly damaged. Two farms had been destroyed, and also a nursery garden. Two engineering plants that made tank chassis, filters for wood generators, de-gassing cans, etc. were destroyed. Other factories had been damaged and a mineral oil factory which supplied oil to the armour industry. A total of 444 local livestock were drowned, including cattle, horses, pigs, goats, sheep, chickens, geese and ducks.

The floods brought much flotsam, mainly building planks, possibly from the barracks of a labour camp. These were collected and resold to people for nominal sums. An eye-witness in Essen said, 'Heele suffered considerable flooding and transport was possible only by boat. Many household items, wardrobes, small cases and similar items, came down the Ruhr.' The arrival of the flood in Essen was expected on the 19th at about mid-day and the town areas of Heele, Werden and Kettioig had to be cleared because of flood damage. The flood water when it came, carried all sorts; dead bodies and all manner of household goods. Two companies of soldiers were ordered to Essen to distribute drinking water in certain areas. The relief work extended to 25th May. In nearly all the Ruhr area, endless lines of water waggons brought drinking water; women and children and old people stood everywhere with buckets or basins, waiting for water.

All the railway and road bridges over the Ruhr were damaged or destroyed, and repair of the Spitzen Power Station which supplied some 140,000 KW daily took some three weeks.

A number of works in the area of Hamm, Dortmund, Witten, Hagen and Gelsenkirchen were all put out of action due to water

(*Top left*) The Möhne Dam the day after the raid
(*Top right*) General Weis at the Möhne Dam the day after the raid

Karl Schutte wearing his Iron Cross

Möhne Dam after the raid

shortage. The gas industries suffered too, and the loss in gas production was 50% for several days which caused rationing of supply at one stage.

A Swedish news reporter wrote:

> The flooding after the dams raid has created great havoc. The town of Soest has for a long time been like an island, and entire buildings swept away. In Dortmund many streets were submerged and traffic restricted to flat bottom boats. The soldiers of the 1914/18 war were saying that not even the gunfire in Flanders had done more destruction than the British attack on the Ruhr.

Frau Noller lived in a village below the Möhne and recalled how she heard the air raid alarm and got out of bed to go to the cellar. It was a clear moonlight night when she heard the great roar of engines as a bomber skimmed over the roof of her house. From them on the noise was indescribable, with the vibrations of the Merlin engines coupled with the bark of gunfire and exploding shells. People in the cellars were soon told to go upstairs in the houses as the water was coming. Before the dam was actually breached, a lot of water had spilled over the edge. Frau Noller managed to reach the roof of a house in a meadow where her brother-in-law lived, when her house was swept away when the flood came.

A communique was issued by the German High Command at 2 pm on the 17th, which read: (picked up by the BBC Monitoring Service)

> Last night a force of Royal Air Force bombers penetrated Reich territory and dropped a small number of high explosives in several places. Two dams were damaged and the subsequent onrush of water caused many casualties among the civilian population. Eight enemy aircraft were shot down.

In Germany, Doctor Josef Goebbels, director of propaganda, issued a news agency report, claiming that the plan of the attack had been thought up by a Jew who had emigrated from Berlin. He also wrote a short news item for the newspapers, particularly the local papers in the flooded areas. The amount of damage caused was published.

What the actual damage caused varies, but the death total was reported as 147 Germans, 712 Russian and Polish workers with 41 more injured in the area of Neheim alone. The Russians and Poles were all buried at Neheim. At the Osterbir cemetery there are 479 Russian and Polish workers buried, plus 40 prisoners of war, also 54 French and a number of Dutch prisoners of war, all apparently held or working in the area affected by the flooding.

On the morning of the 17th, two Spitfires, flown by Flying Officer Searle (in EN346) and Flying Officer Efford (EN141), of 542 Photographic Reconnaissance Squadron, flew over the area and took photographs from 29,000 feet. Water was still flowing through the great gap in the Möhne dam and the photos showed the flood had reached a point some sixteen miles down the valley. Two villages were under water, bridges had been swept away, power stations and waterworks isolated and railway communications disrupted. Flying Officer Efford saw mud deposits west of the Eder along the Eder valley for up to twenty miles. At the Eder the generator house, the switch park of the Bringhausen Storage Power Station, was seen to be flooded.

By the late afternoon most of the water had drained away but the switch and transformer park showed signs of silting, a portion of the northern part being washed away. One of the supports of the Herdecke Viaduct on the direct route from Dortmund to Hagen and Düsseldorf was broken, leaving the line suspended precariously over the gap. Disruption to barge traffic on the Rhine was considerable, at one point the delay in the journey from Cologne to Holland amounted to six days. At the Sorpe, water was seen to be running down the face and carrying earth from the dam into the compensating basin.

On the 18th a further recce was made by 542 Squadron, the sortie flown by Flying Officer Scott (EN141). His photographs and report found the lakes mostly drained away, and hangars and barracks, ammunition dumps and landing ground at Fritzler

Eder Dam breached

Eder Dam today. Note two outlets missing on the left

airfield submerged. It also showed that the Ennepe dam had a minor breach, so Bill Townsend's trip had not been in vain.

On the 16th, the water in the Möhne lake had been 132 million cubic feet. On the 18th it was down to a mere 14.4 million. The Sorpe dam water level had dropped to a position coinciding with the bottom of the crater which was identified as that having been caused by the bomb. A portion of the casing, made of steel 5/16ths thick, was picked up at the bottom of the crater. The Möhne and the Sorpe later contained little more than one third of their maximum capacity. Over 3,700 million cubic feet of water was lost from the Möhne, and at the Eder, 4,272 million cubic feet lost. The reservoir water for the whole district then stood at a figure substantially below one year's full consumption and had it not been for the destruction of a large part of the Ruhr's industry, the water shortage would no doubt have been acute.

When Tony Burcher had been picked up and taken off to the local police station, he was laid out on a wooden bed to help with his broken back. He had developed a craving thirst, and while waiting for the arrival of a promised doctor, he called out to the German Corporal guarding him. 'I want a drink of water.'

'You want *wasser*?' replied the guard.

'Yes, *wasser*, please,' gasped Burcher, and the guard went out, then returned.

'So, you want a drink of *wasser*, do you; thanks to you and your comrades there is no *wasser* any more!' With a grin on his face Burcher realised that they had done their job after all. In his pocket was the piece of stone given to him by the little boy to drop on the Germans. It had saved him again.

CHAPTER TEN

Eleven came home …

As ordered, the ground crews of 617 Squadron reported for duty at 8 am on the morning of 17th May. They began work on the Lancasters when suddenly the Tannoy system broke into life and all 617 Squadron personnel were ordered to report to the airmen's mess at 10 o'clock.

The mess was packed with all ranks and trades. They were called to order and Wing Commander Guy Gibson addressed them dramatically.

'Ladies and Gentlemen. Last night we went out and altered the map.' He then gave them a brief account of the raid and the targets, then thanked them all and told them to proceed to the guard room in an orderly manner, where they would be all given three days leave. Aircrew received seven days.

On that morning Squadron Leader Taylor of 49 Squadron, was sent to RAF Scampton for Link trainer instruction. As he arrived at the main gate he saw a line of airmen and WAAFs stretched across the main road, stopping all traffic; they were hitching lifts to the north or south into Lincoln. On turning into the camp he saw a large gaggle of bods standing outside the guardroom, and service police handing out leave passes. A name would be called out and a voice would respond, then a pass would be passed from hand to hand to the lucky person. He would then join the rest out in the road. After an hour at Scampton, Taylor returned to his squadron at Fiskerton where he discovered the reason for all the excitement.

Gibson did not go immediately. He went off to the Orderly Room to compose the letters to the next-of-kin of those who had failed to return. Sergeant Heveron assisted him with this sad task.

A message had been sent by Sir Arthur Harris to AOC 5 Group, Ralph Cochrane, that read:

Please convey my warmest congratulations on the brilliant and successful execution of last night's operation to the aircrew. I must say that what they have been through in training and their skill and determination in pressing home their attack will for ever be unsurpassed in the Royal Air Force. In this memorable operation they have won a major victory in the Battle of the Ruhr, the effect of which will last until the Boche are swept away in the flood of final disaster.

The first news of the operation was given by Sir Archibald Sinclair, Secretary of State for Air, when speaking next day at the Norwegian Independence Day celebration at the Royal Albert Hall, the King of Norway, King Haakon, and the Crown Prince were present. A telegram was also sent to HQ Bomber Command by Sir Archibald to congratulate Gibson and his squadron on behalf of the War Cabinet on the great success achieved. It continued, 'This attack, pressed home in the face of strong resistance, is a testimony alike to the tactical resource and energy of those who planned it, to the gallantry and determination of the aircrews and to the excellence of British design and workmanship.' The Cabinet also noted with satisfaction, the damage inflicted on German war power.

The morning after the operation, Group Captain Verity was at the Air Ministry building in King Charles Street, off Whitehall, discussing future operations with Air Vice-Marshal Sydney Bufton. Verity had received the signal concerning the success of the raid. Suddenly the door opened and there stood Barnes Wallis, unshaven and dishevelled, having just returned from Scampton. He had spent best part of the night there, the other part at 5 Group HQ; he was beside himself with excitement. Within ten minutes a despatch rider came in from RAF Benson with the first reconnaissance pictures taken of the breached dams, which gave complete evidence of the success of the operation. After months of planning and setbacks, the three of them were terribly excited. A short while later the telephone rang; it was Sir Charles Portal. He had heard Wallis was there and wanted to see him at once, so off he went, his hands full of photographs, signals and all manner of bits and pieces.

Verity felt strongly that the crews of 617 had contributed 90% to the success of the operation with their brilliance and bravery, without which the planning was worth very little. They were the real heroes in the best sense of the word and had made a tremendous sacrifice.

One funny thing happened, recalls Verity, some days later, following the press announcement of the raid. Air Ministry were showered with letters sent to congratulate all who took part, and many suggestions were made by would-be 'dam busters', concerning other targets that could be attacked. Among them were several postcards showing the picture of a dam in southern Germany which completely mystified him, as he thought all the dams had been covered in their research and this was not one of them. It transpired that the dam was on the road from Munich to Oberammergau which was near a little café where travellers could stop for refreshment. At the café postcards of the dam could be bought but it was quite unimportant.

In Germany on the morning of the 17th, Karl Schutte was still on duty on the Möhne dam.

'In the grey morning an observer plane passed over and at which we fired at immediately. The firing was good but it was not possible to shoot him down. In the meantime it got lighter and we could begin our work, of which there was enough. First the gun, and then the accommodation which was like a rubbish dump. Beams broken, beds and wardrobes riddled with shot. Stairs collapsed, ceilings ruptured and so on. After several hours there was the arrival of our General by Fieseler Storch (light aeroplane). In the afternoon the presentation of Iron Crosses, as well as the arrival of several flak units. On the following day a visit from the chief of all air force units, Generaloberst Weise. His redeeming words acted as a calming influence and because of the one plane we did shoot down, we thought we had done our best. We all felt very proud that we were part of the attack.'

Back in England the congratulation continued. Lord Trenchard sent his congratulations: 'The wonderful work of Bomber Command is being recognised by all now.' Sir Charles Portal sent particular congratulations to Sir Arthur Harris. Although against the idea earlier, Sir Arthur realised the weapon had been proved and sent a telegram to Barnes Wallis:

But for your knowledge, skill and persistence, often in the face of discontentment and disappointment, the efforts of our gallant crews would have been in vain. We in Bomber Command in particular, and in the united nations as a whole, owe everything to you in the first place for the outstanding success achieved.

Sir Arthur was summoned to Buckingham Palace by the King, where His Majesty expressed his personal congratulations on the Command's recent exploits, particularly on the outstanding success of the raid on the dams.

On 27th May the King and Queen visited RAF Scampton. They arrived by road at 1 pm and went straight to the Officers' Mess for lunch. Here photographs were taken. Many of the pilots who had flown on the raid were in the group with Group Captain Whitworth and Guy Gibson. At 2 o'clock they drove to the tarmac where crews of 617 and 57 Squadrons and the station WAAFs were formed up. One aircraft of 617 and one of 57 were on display along with air and groundcrews.

Barnes Wallis was also present, and the Royal couple spoke to him and to many of the airmen, including David Maltby, Micky Martin, Les Munro, Ken Brown and Joe McCarthy. At 2.30 they all went into No 2 Hangar's crewroom (still there today) where they were shown the models used for the preparation of the raid, also reconnaissance photographs of the Möhne, Eder and Sorpe dams through stereoscope lenses. These showed the flooding of the valleys some fifteen miles from the breaches and also the smashed dams themselves. Kassel was also to be seen, with flood water filling the Tiergarten and important factory areas.

Now that the squadron had established itself, they had designed two Coats of Arms. Gibson submitted them both to His Majesty, one showing a hammer parting the chains on the wrists of Europe, with the motto – 'All the Map', the second which Gibson himself favoured, showed a breached dam in the centre with the motto –

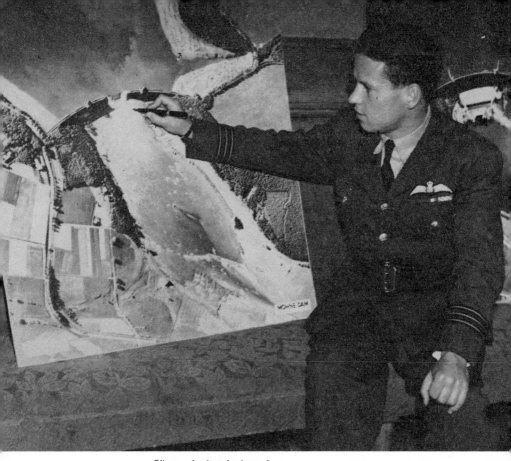

Gibson signing the breach

The Men Who Breached the Dams, Scampton, July 1943

'Après moi le Déluge' (After me the flood). No decision was made by the King, who suggested the expert advice of the Chester Herald be sought.

The Royal Party were then invited to inspect the loaded aircraft in the hangar, where Wallis explained his bomb and how it worked. They then went to the bomb dump where the armament section and WAAF drivers were lined up. At 3 pm the King inspected the station workshops and equipment stores before they took their leave at 3.30.

Harlo Taerum was thrilled when the King and Queen were introduced to him and said afterwards, 'The Queen was most charming and gracious, it was really quite a day.'

For the men who had survived the raid on the dams, this was just the start of a period of popularity which put them in the forefront of public life. The whole idea of the raid captured the imagination of the population in war-torn Britain, where a victory, any victory, was worth celebrating. That this handful of flyers had carried out such an unusual attack upon Germany with the success achieved was worth reading about, and talking about. In high places too, it was quickly felt that just to congratulate these men was not enough. Some more tangible evidence of a people's thanks should be given, their courage recognised.

It was obvious that a large portion of decorations must be awarded to these men, and made quickly. While on leave, Gibson's navigator Harlo Taerum, went to London, as did most of the Australian and Canadian crew members, the British men normally being able to visit their families. Taerum was awoken one morning to be told he had been awarded the DFC and a little time that morning was spent sewing on the ribbon of this coveted decoration to his uniform.

Guy Gibson's recommendation for the highest award, the Victoria Cross, was made on the 19th May and approved by the King on the 22nd. Tammy Simpson of Mick Martin's crew spent some of his leave with friends in Norwich but went to London on his roundabout way back to Scampton. He was told of his DFM award on 25th May, the day he arrived in the city. Mick Martin was awarded the DSO and Jack Leggo and Bob Hay, bars to their

DFCs. They went for lunch at Shepleys, joined the 10 Club and spent the night at the Strand Palace Hotel, leaving for Scampton the next morning.

The whole of Gibson's crew were decorated. Sergeant John Pulford received the DFM. He came from Hull and had been a motor mechanic before the war and both he and his brother Thomas flew in the war. On the night of being awarded his decoration he and his brother went out to celebrate, discarding their uniforms for civilian dress for a change. In a pub they both had white feathers put in their pockets which upset both men and they never again went out without wearing their uniforms.

George Deering, front gunner who had used his guns to advantage on the outward trip as well as over the Möhne, received the DFC. He was born in Ireland of Scottish parents but was educated in England before going to Canada where he enlisted in the RCAF in 1940. He was commissioned on 18th May 1943. Robert Hutchinson was 25 and from Liverpool. With 617 he was also the Signals Officer and made history by sending the message 'Nigger' back to Group HQ from the Möhne. He received a bar to his DFC earned with Gibson in 106 Squadron.

Spam Spafford DFM added a DFC to his collection for his accurate placing of the bomb against the dam while Trevor-Roper, the rear-gunner, whose twenty-eighth birthday fell just two days after the raid, collected a bar to his DFC.

Sadly all six men were later killed. Pulford died in a crash on 13th February 1944 near Chichester, Sussex, Deering, Taerum, Hutchinson and Spam Spafford were all lost in a raid by 617, on the infamous Dortmund-Ems Canal on 16th September 1943, their Lancaster being piloted by the new CO of the squadron, Wing Commander G. Holden DSO & bar, DFC, who also died. Trevor-Roper was lost flying with another crew on a raid on Nuremberg on 31st March 1944, while with 97 Squadron. The Lancaster in which he was rear gunner was shot down by Major Martin Drewes of III/NJG 1, the German's sixty-eighth night victory (of a total of 96). When the body of 'Trev' was found it was still in the turret.

Micky Martin received the DSO. He had made his attack only moments after seeing Hopgood shot down by flak from the dam. He went on to receive two bars to his DFC the first while still flying with 617 Squadron, the second as a night fighter pilot with 515 Squadron in 1944. He was also awarded a bar to his DSO with 617. After the war he received the Air Force Cross for leading the Vampire Squadron No 54 in a Mosquito for the first crossing of the Atlantic by jet aircraft. In 1947 he received the Britannia Trophy and the Oswald Watt Memorial medal for a record-breaking flight to Cape Town in a Mosquito, both awards were for the most outstanding contribution to aviation in 1947. Remaining in the RAF he became a CB in 1968, then Knighted (KCB) in 1971, retiring as an Air Marshal in 1974. He is the President of the Bomber Command Association.

Most of his crew were also decorated. Navigator Jack Leggo who already had a DFC following a tour with 50 Squadron in 1942, received a bar to this decoration. He came from Sydney, Australia and had the job of training all the unit's navigators to the high standard required for the raid. Len Chambers, Martin's 24-year-old New Zealand Wireless Operator, received the DFC, having already flown with 75 (NZ) Squadron. Bob Hay DFC was thirty-one, from Gawlor, Australia and became 617's Bombing Leader. He too had a tour behind him, flown with 50 Squadron. He received a bar to his DFC. Tom Simpson was also Australian, from Hobart, Tasmania. He collected the DFM as rear-gunner. He already had a recommendation in the pipe-line for his tour with 50 Squadron, but this was cancelled in favour of the 617 Squadron award.

The remaining two crew members were engineer Pilot Officer Ivan Whittaker, another ex-50 Squadron man, and Pilot Officer Bert Foxlee DFM from Queensland, Australia, also ex-50 Squadron, the front gunner. Whittaker went on to win the DFC and bar with 617 and became a Group Captain in 1965. He died in RAF Halton hospital in 1979. Foxlee added a DFC to his DFM in 1944, while still flying with 617. All survived the war except Bob Hay who was killed in action with 617 on 12th February 1944 on a raid on Anthear Viaduct.

David Maltby also received the DSO as the pilot of the Lancaster whose bomb was actually seen to make the breach in the Möhne. He was very upset after the raid at the number of men who had failed to return, but said that the sight of the dam breaking had

L to R: Onacia, Sutherland, O'Brien, Brown, Weeks, Thrasher, Deering, Radcliffe, McLean, McCarthy, McDonald
Front L to R: Pigeon, Taerum, Walker, Gowrie, Rodger

(*Left*) David Shannon, Charles Whitworth, The King. (*Centre*) Micky Martin, Charles Whitworth, The King, Guy Gibson. (*Right*) Joe McCarthy, The King, Guy Gibson

(*Left*) Les Knight, The King. (*Centre*) David Maltby, Charles Whitworth, Ralph Cochrane, Barnes Wallis, The King. (*Right*) Ken Brown, Charles Whitworth, The King.

been fantastic!

Maltby's navigator, Flight Sergeant Vivian Nicholson, won the DFM. He was twenty, came from County Durham and the raid on the Möhne had been his first operational sortie!! Pilot Officer John Fort, the bomb-aimer, received the DFC. He came from York and was on his second trip, having flown just one op with 97 Squadron! Maltby had picked up this crew when he arrived on 97 in March 1943 to commence his second tour. Sergeant Bill Hatton, another Yorkshireman, was his engineer, Flight Sergeant Tony Stone, from Winchester was wireless operator, Flight Sergeant Victor Hill, front-gunner (he had flown with 9 Squadron before joining 617), and rear-gunner had been Sergeant Howard Simmonds, from Burgess Hill, Sussex.

They were all killed on 14th September 1943 when recalled from a proposed attack upon the Dortmund-Ems Canal. In turning back, Maltby hit the sea with a wing-tip and went in. Dave Shannon circled the area for some time but all were lost. The date of Maltby's death is sometimes given as 18th September but this is the date his body was picked up by Air Sea Rescue.

The DSO also went to Dave Shannon for his attack on the Eder. He stayed with 617 Squadron, flying nearly seventy operations before being taken off ops. With 617 he added a bar to his DFC in 1943 and a bar to his DSO in 1944. He is now a successful businessman in London.

His navigator, Flying Officer Dan Walker DFC, RCAF from Alberta, received a bar to his medal. He remained in the RCAF after the war, retiring as a wing commander in 1967. Brian Goodale DFC, aged twenty-three, had flown earlier with 51 Squadron 1941-42 and received a bar to his DFC for his part in the raid. As wireless operator he later became 617's signals leader. He retired from the RAF in 1967 as a squadron leader and died in a Cambridge Hospital after a long illness, in 1977.

Flight Sergeant Len Sumpter, Shannon's bomb-aimer, aged thirty-two from Kettering, Northants, won the DFM. He had been in the Grenadier Guards from 1928 to 1931 and again in 1939 before transferring to the RAF. He flew thirteen ops with 57 Squadron, having been wounded on one occasion, before joining 617. Commissioned in June 1943, he later won the DFC while still with the squadron in 1944 (Mosquito marker), and he returned for another tour with 617's aircraft in 1945. Jack Buckley, from Bradford, was Shannon's rear-gunner and had flown with 75 Squadron 1941-42. He received the DFC. In August 1943 he became 617's gunnery leader.

Les Knight was another pilot who received the DSO for his attack on the Eder dam. He had been recommended for the DFC for operations with 50 Squadron but this was deleted in favour of the DSO award with 617. Knight's navigator, Flying Officer Harold Hobday, from Croydon, Surrey, received the DFC. He too had been recommended for the DFC for his tour with 50 Squadron, but this was cancelled in favour of the 617 award.

The fifth pilot to receive the coveted DSO was Joe McCarthy. He attacked the Sorpe despite difficulty in locating the target. He went on to win a bar to his DFC with 617 Squadron, completing around fifty trips on bombers.

McCarthy's Canadian navigator, Don McLean, came from Toronto and had been a school teacher in North Bay, Ontario. He came to 617 from 97 Squadron where he was known as the 'Gee Box Wizard'. He ended his war having flown a total of fifty-two operational bombing raids, returned to teaching but rejoined the RCAF, retiring a wing commander.

Sergeant George Johnson, thirty-two, came from Horncastle, Lincolnshire, but later lived near Nottingham. Failing to become a pilot he remustered as an air gunner then a bomb aimer. Flying with 97 Squadron he completed twenty-eight ops before joining 617. He won the DFM for his drop on the Sorpe after nearly an hour in the area of the dam. He retired from the RAF in 1963 as a squadron leader.

Bill Townsend DFM was awarded the Conspicuous Gallantry Medal for his attack on the Ennepe dam. His navigator won the DFC. This was Pilot Officer Lance Howard aged twenty, from South Freemantle, Western Australia. He had flown twenty-five ops with 49 Squadron before joining 617. Flight Sergeant George Chalmers from Harrogate, Yorkshire, flew as wireless operator and received the DFM. His DFC following a commission, came for continued service with the squadron and a total of sixty-five missions.

Townsend's bomb aimer, Sergeant Charles Franklin DFM, was

awarded a bar to his decoration, which had only been awarded a few weeks beforehand, for his tour with 49 Squadron. This proved to be the only bar to a DFM awarded to any member of 617 during the war. Aged thirty, he came from Dagenham in Essex. He died on 25th January 1975 in Birmingham.

Both air gunners won DFMs. Both were twenty-two and both had flown ops with 49 Squadron. They were Doug Webb from Leytonstone, Essex, and Ray Wilkinson from South Shields, County Durham. Doug returned to fly a second tour with 617 in 1945 and so did Ray. Both survived the war; Doug lives in Surrey but Ray died in 1980.

Ken Brown also received the CGM for his attack on the Sorpe, despite adverse conditions. He remained in the RCAF after the war, taking his English wife back to Canada with him. He retired in 1967 with the rank of squadron leader.

His navigator, Sergeant Dudley Heal, from Hampshire, received the DFM, having just started a tour with 44 Squadron. He survived the war, becoming a senior customs officer. The bomb aimer, Sergeant Stefan Oancia, from Stonehenge, Saskatoon, Canada, aged twenty, also received the DFM and now lives in America.

On 21st June 1943, the members of 617 who had taken part in the dams raid, set off for London. Thirty-seven of them were to be decorated the next day at Buckingham Palace. Their train left Lincoln at 2.20 pm, the men travelling in a specially reserved carriage. In all, the party numbered about forty-five. As one can imagine, the journey was not without incident. At Grantham, Tammy Simpson wanted to exchange his seat for the cab in the engine. On arrival in London they set off for the Strand Palace Hotel, and, some, for whom the journey had been too much, went straight to bed.

The big day – 22nd June. ETA at Buckingham Palace was set for 10.15 am sharp. Best uniforms was the order of the day. Each man had been given two tickets for relations or friends.

They arrived on time and soon all were assembled and awaiting the arrival of His Majesty, but to their surprise it was the Queen who appeared. With her was Lord Clarendon, the Lord Chamberlain, Lady Delia Peel and other members of the Royal

Household. They all walked to the dais while the Guards band played the National Anthem. This was the first time since the reign of Queen Victoria, that a Queen had presided over an investitute, the King being away at the time, on a tour of North West Africa and Malta.

The first to be decorated was Wing Commander Guy Gibson DSO, DFC. It was normal on these occasions for recipients of the Victoria Cross to be presented last, but on this special day he was the first to be decorated. With his VC, Gibson also received the bar to his DSO awarded to him earlier, making him, at twenty-four, the most highly decorated member of the Royal Air Force in WW2 at that time. The Lord Chamberlain read out the citation, describing the breaching of the dams as one of the most devastating attacks of the war.

Her Majesty said how sorry the King was that he could not be present. Looking at the line of Gibson's men, she said, 'Have you brought all your fellow raiders with you?''

'They are all here, Ma'am,' he replied, although in fact John Pulford was sick – he received his DFM from the King on 16th November 1943.

Tammy Simpson recorded in his diary what a wonderful thrill it had been speaking to the Queen as she pinned the DFM medal to his tunic. Douglas Webb also remembers the occasion, the Queen asking him if any of the men were suffering from any ill effects of the raid.

Roy Chadwick was also decorated. As designer of the Lancaster bomber which had taken Wallis' bomb to the dams, he received special mention and became a Commander of the Order of the British Empire. He said that he was proud to have received the honour and to have been presented with it alongside the boys who had flown on the raid.

Many photographs were taken outside the Palace. Some were official group photographs, others, such as those taken with family or friends, being not so official. In the evening, Messrs A.V. Roe & Co, Ltd, makers of the Lancaster, gave a dinner for the squadron, held at the Hungarian Rooms, Dorland House, Lower Regent Street, in London. The menu cards carried the legend – 'Dam Busters'.

This was the first occasion they were so called, and to this day they remain known by this title, the name going down in history.

The meal consisted of Crab Cocktail, Crème Santé, Caneton Farci à l'anglaise, Petit Pois à Menthe, Pommes Nouvelles, Asperges Vertes, Sauce Hollandaise, and Fraises au Marasquin.

Jack Hylton and his band provided the entertainment, together with his singer, Elsie Carlisle. It was an impromptu show, for they had been dining upstairs with Arthur Askey, when some of the 617 boys came to ask for autographs. They in turn asked the 617 boys if they could join them! They did, putting on a show for the squadron that Arthur Askey remembers as going down like a bomb!

During the evening, Gibson signed a large photograph of the breached Möhne dam, and his signature can be seen right in the hole of the breach. The rest of 617 signed the photograph on the right-hand side of the breach. This was presented to the 'backroom boy', with the compliments of the Ministry of Aircraft Production. Although the name of this 'boy' was withheld for security reasons, this was Barnes Wallis. Gibson himself was presented with a model of a Lancaster bomber in silver, by Avros.

Tammy Simpson recorded in his diary, 'Dinner given by A.V. Roe people, marvellous show, great people.' Jim Heveron remembers it cost sixpence to place his hat in the cloakroom, but with the usual service quick thinking, a gang of chaps put their caps in a suitcase and only paid 6d for the one case!

Gibson was soon in great demand for public appearances. Towns launching savings drives requested him to attend. When he could, he did so. Being a totally new role for him, he was often asked what his usual line of speech might be. He usually replied, 'I'm not tactful, I tell them the truth. I tell them that whatever the papers say, there is nothing in the war a civilian can do that will equal the sacrifice asked of the bomber crews who go out to Germany night after night losing about thirty each time, until their time comes. I tell them there's nothing they can do to equal that, but that they can try by giving money. It's uncomplimentary, but it works.'

At Maidstone in Kent, they wanted to raise around £750,000, Gibson being the main attraction. In fact they raised £1,750,000. He said on that occasion:

'A man is not normal if he's not frightened. Even after dozens of raids there's a tightening up of the stomach as you approach the target, and you realise in the next ten minutes you'll either be alive or dead. It's like going into a cauldron, searchlights for a ring of your cauldron. The incendiaries are the bubbling brew, the flak and the spent flak smoke are the sparks and steam. Only when you're in the cauldron does the fear depart entirely and your training takes over. It's the difference between waiting in the wings to go on, and then doing a well-rehearsed act. If you drop your bombs accurately, there's a terrific feeling of elation, you feel the plane is lighter and more buoyant.'

He knew his trade. He had now flown some seventy-four night bombing raids over enemy territory plus ninety-nine fighter sorties since the war began. His aircraft had dropped an estimated 370,000 pounds of bombs on German and Italian targets. He said:

'The success of the attack on the dams was due to hundreds of technicians and above all to the Air Officer Commanding the Group, and the Senior Staff Officer. We flying crews are indebted to them.'

For his leave after the dams raid, he took his wife Eve to Cornwall, his old boyhood stamping ground. He could get away from it all by walking in the countryside of his youth, or taking out a 30-foot fishing smack with an old sea-faring man, Dick Perkin. They would start out at 5 am, cast their lines, mark them with a float, then go back at noon to haul in the catch. One day they pulled in a large skate, a huge monk fish, and a 12 foot shark! For a while he could relax. Relax and leave the war behind.

CHAPTER ELEVEN

... Eight of our aircraft failed to return

Eight Avro Lancaster bombers and fifty-six members of aircrew failed to return from the attack on the German dams. All had fallen, flying at low level, with so little time to recover, following hits by anti-aircraft fire or from striking objects, that death had resulted almost instantly. This was the danger of flying operationally at low level, especially at night, whether in moonlight conditions or not.

Although flak positions had only a relatively short time to fire at a low flying aeroplane, it could present a good, if fleeting target in moonlight over a flat countryside.

The first crew to go down had been that of Sergeant Vernon Byers, the 32-year-old Canadian from Saskatchewan. He had taken off at 9.30 pm as part of the second group and fell one hour and 27 minutes later. He had just reached the enemy coast when he came under AA fire from the heavily defended island of Texel. Geoff Rice saw Byers' Lancaster hit by the gunfire when it was at 300 feet. The bomber veered off course and crashed into the sea near Vlieland Island off the Frisian chain.

A message received at 10.50 pm by the local authorities of the town of Den Helder in the north of Holland, stated that a plane had crashed some miles east of the town between Texel and the Afsluitdijk (the large dyke that connects both parts of northern Holland). Byers and his crew perished. He was remembered by the Province of Saskatchewan Department of Natural Resources, when it named a bay, Byers Bay, in 1968.

Byers' flight engineer, Sergeant Alastair James Taylor, was twenty-two, and came from Morayshire and became a Halton 'brat' in 1939 before volunteering for aircrew duties. Later going to 467 Squadron in February 1943 he began operations. After a leaflet raid on St Nazaire, his aircraft was diverted to Kinloss a few miles from his home in Elgin. Arriving at 2.30 in the morning he borrowed a bike and rode home to spend a few hours with his parents. This proved to be the last time they were to see him. He had only flown four trips when he was killed. During the training period for the dams, his pilot often flew low over Taylor's home. His brother, now a farmer in Canada, was also a flight engineer in the RAF and completed a tour in Bomber Command.

The navigator in Byers' crew was Pilot Officer Jim Warner who had come up through the ranks, from leading aircraftman in 1940 to commissioned officer in 1942 (on the Runnymede Memorial, Panel 130, he is shown as a flying officer). Sergeant John Wilkinson was the wireless man, from Antobus, Cheshire, while the bomb aimer was Sergeant Arthur Neville Whitaker (his commission came through on 18th May!), had enlisted on 3rd September 1939, volunteering for aircrew in 1941. The two gunners were Sergeant Charles Jarvis from Glasgow, aged twenty-one, and Sergeant James McDowell.

The sea only gave up the body of McDowell, found by a vessel on 22nd June, in a part of the Waddensea known as Vliehous, near Beacon 2. He was buried in Harlington General Cemetery in the Netherlands, Plot E, Row 4, grave 11. The rest are remembered on the Runnymede Memorial.

The second casualty was Flight Lieutenant Bill Astell DFC, who survived the operations in the Middle East only to die when his Lancaster hit a high voltage cable near Dorsten aerodrome near Marbeck, Germany, on the outskirts of Borken. He and his crew, Sergeant John Kinnear from Fife, Scotland, Pilot Officer Floyd Wile from Nova Scotia, brother Canadians Sergeant Abram Harshowitz and Sergeant Francis Garbes, both from Ontario, Flying Officer Donald 'Hoppy' Hopkinson from Oldham, Lancs, and Sergeant Richard Bolitho, from County Antrim, Northern Ireland and later Nottingham, were all killed instantly. They were all buried in Borken, but after the war were re-buried at Reichswald war Cemetery at Cleve. When John Kinnear's sister and brothers visited Reichswald afterwards, they all said they had never experienced such tranquillity and peace, and could not wish for a better resting place for anyone.

Also at Reichswald are the crew of E-Easy, piloted by Flight Lieutenant Robert Barlow. He survived Astell by some thirteen minutes and like Astell, he too hit a cable. Barlow, the first pilot to take off for the raid, got as far as the south-end of Emmerich before meeting disaster. All his crew were killed instantly with him. Sergeant Sam Whillis, aged thirty-one, a married man from Newcastle on Tyne, Flying Officer Phillip Burgess aged twenty, Flying Officer Charles Williams DFC, a Queenslander aged thirty-four, who had a tour with 61 Squadron under his belt, Sergeant Alan Gillespie DFM, a former solicitor's clerk from Carlisle, who had just completed a 33-op tour with 61 Squadron, and gunners Flying Officer Harvey Glinz, twenty-two from Winnipeg, and Sergeant Jack Liddell from Weston-super-mare.

The report of the RAF's Missing Research and Enquiry Unit, mentions the local people in the area of Barlow's crashed aircraft at Heeren-Herken-Haldern, south-east of Emmerich, talking of a large bomb found in the wreckage which was later removed after it was defused, for research by the Luftwaffe. The report states that the bomb was 3,900 kilograms (8,580 pounds) and was of a type used to breach the dams. A photograph of this bomb was taken by Herr Wolter, leader of the railway police in the Reichsbahn-Direction Wuppertal (District Railway), was thought to have been taken at Düsseldorf-Kalkum, the base of the bomb disposal squad. When found, it had been buried 50 cm into the ground.

German scientists and German armament engineers later inspected the bomb, but they had no conception of the bouncing bomb principle.

Loss number four was the gallant Hopgood, the gentle man who felt compelled to fight in the war but had no wish to see the suffering that war brings. Just how badly he had been wounded when his Lancaster was hit soon after reaching the enemy coast, we will never know. But there was not the slightest hesitation in going on. With himself wounded, his wireless operator and rear gunner also wounded and his front gunner possibly dead, he could, with honour, have turned back. His and his crew's training dictated otherwise. They went on, only to perish shortly after dropping their bomb.

Only Pilot Officer J.W. Fraser, his Canadian bomb aimer, and Tony Burcher in the rear turret survived. Sergeant Charles Brennan, the engineer, who had flown a tour with 106 Squadron, was killed. Canadian Flying Officer Ken Earnshaw from Alberta, a married man who skied the three or four miles in winter to the school where he taught, was killed. He had flown a tour with 50 Squadron and his loss was a bitter blow to his school, and a great disappointment as they had been told that having finished his tour he was soon to be back with them in Canada. Sergeant John Minchin, the wounded wireless operator, aged twenty-six and married, a tour-expired veteran from 49 Squadron, was killed. Pilot Officer George Gregory DFM, from Glasgow, was killed. He had enlisted in the RAF in 1939 and had flown a tour with 44 Squadron in 1941-42. He had won the DFM as a sergeant in 1941 having shot down a Messerschmitt 110 night-fighter on one occasion and on another had directed his pilot in evading attacks by three enemy fighters. His wife Margaret had been with him at Buckingham Palace when he received his decoration from the King and she lived near Scampton when he was with 617. Guy Gibson telephoned her upon his return from the Möhne, telling her of her husband's possible loss. She last saw George at Lincoln station when she was going to visit her mother who was ill.

Siegfried Reinhold, the policeman's son, who was later to help in rescue work at the church at Hemelfonten, remembers that his father had to go to the Möhne, returning the next day to say, 'It's all over, the water's off.' Siegfried had gone to school later, hearing of the Lancaster that had been shot down. Four or five men had died in the bomber and two had been taken prisoner. The man who took one of the men prisoner was Fris Yahbert, a 55-year-old teacher and part-time policeman. The man he took to the local gaol was Fraser. Burcher was brought in by members of the Hitler Youth. The dead were taken to the local Luftwaffe base at Werl, by Alpine troops, then taken to Soest for burial. One 12-year-old boy visited the crash site and picked up a piece of triangular wood. Being charred and burst he threw it away, but it had been the bomb-sight used by Fraser to line up on the Möhne. The boy kept, however, the optical bomb sight from the Lancaster.

The five dead flyers were later reburied in Rheinberg War

Cemetery, north of Krefeld. Their courage had deserved a better fate.

The fifth to go down came from the third reserve-group, Pilot Officer Lewis Burpee DFM, who loved music and reading. His wife, Lillian, lived in two rooms near the airfield at Scampton and when Lewis's death was confirmed, she left England for Canada in July 1943 to live with his parents. There her son, Lewis J. Burpee the second, was born on Christmas Eve.

Dazzled by searchlights, Burpee had hit some trees and crashed on Gilze Reijen airfield at 1.53 am, dying with his crew when their bomb blew up. His flight engineer had been Sergeant Guy Pegler from Bath, Somerset, aged twenty-one, who had been a member of ground crew before volunteering for aircrew duties. He later flew with 106 Squadron. Tom Jaye came from County Durham, aged twenty. Too short for a pilot, he became a navigator and also flew with 106 Squadron, having trained with Pan American Airways in Miami, Florida. Pilot Officer Len Weller came from Harrow, Middlesex, and had been another former 106 Squadron man. He was twenty-seven years old. The bomb aimer was Sergeant James Arthur from Ontario, the son of a parson, having joined the RCAF in 1941. The two gunners had been Sergeant Bill Long from Bournemouth and Flight Sergeant Joe Brady, another Canadian from Alberta. All were initially buried at Prinsenlage, two miles west of Bread, in Holland. In 1948 they were reburied in Bergen-op-Zoom Cemetery, Holland.

The second Lancaster of the third group to go down, fell at 2.35 am, piloted by Pilot Officer Warner Ottley DFC, shot down by AA fire north of Hamm. There was a photograph of the crashed bomber published, the caption underneath reading: 'The end of a murder plane. This Lancaster bomber will not return to England after its slight intrusion against German territory.' The crash occurred at Boselagerschen Wald in Heessen, north-east of Route 63, between Hardinghauser and Herrensteiner Knapp hills, north of Hamm. He and his crew were also buried in Reichswald Cemetery. Sergeant Ronald Marsden, twenty-two, from Redcar, Yorkshire was the engineer; Flying Officer Jack Barrett DFC, twenty-one, from Essex, navigator, who had won his DFC with 207 Squadron, but did not live to know it, being killed before it was announced. Sergeant Jack Guterman DFM, twenty-two, from Guildford, Surrey, the wireless operator, had flown fifty-four trips, thirty-eight as mid-upper gunner and sixteen as WOP, all with 207 Squadron. He was on his second tour when the call came for him to join 617.

Flight Sergeant Tommy Johnston came from Bellshill, Lanarkshire, and had also been with 207 in 1942. He was the youngest of three brothers and one sister. The two gunners were Flight Sergeant Freddie Tees and Sergeant Harry Strange. Strange came from London and was twenty, but had already flown ops with 207 Squadron. Freddie Tees was the only one to survive the crash; he was found alive but burned having been thrown from the rear turret and picked up four kilometres from the crash. He was taken to hospital and finally to a POW Camp No L6, at Heydekruge.

The seventh casualty had been that of the B Flight commander, Squadron Leader Henry Maudsley, shot down at approximately the same time as Ottley. Damaged while attacking the Eder, Maudsley flew the crippled Lancaster as far as Emmerich (were Barlow had gone down), where it was hit by AA fire. It crashed at Klein-Netterden. Maudsley and his crew were taken to Düsseldorf for burial. They were: Sergeant John Marriott DFM who had just completed a tour with 50 Squadron to win his DFM. His tour included a daylight op to Le Creusot in October 1942 and another daylight trip to Milan a week later. His two sisters received his DFM from the King at Buckingham Palace. Flying Officer Robert Urquhart DFC, the navigator, was from Canada. He too was tour expired from 50 Squadron, and his DFC was in the pipe-line when he was killed. On 17th December 1942, he had been wounded by flak during a low level raid on Soltau. The wireless operator, Sergeant Allan Cottam, aged thirty, was also Canadian, from Alberta, and came to 617 from 50 Squadron. Pilot Officer Mike Fuller, the bomb aimer, was from Kent and had flown with 106 and 50 Squadrons as an NCO, being commissioned in March 1943. He was twenty-two. The gunners were Flying Officer Bill Tytherleigh DFC, another whose DFC was coming through following a tour of forty-two ops, ending up as squadron gunnery leader. He came from Hove, Sussex, and was twenty. The other was Sergeant Norman Burrows who had also flown with 50 Squadron. All seven men were later reburied at Reichswald.

There is a bronze plaque to Maudsley, about two feet in diameter, in the church at Sherborne, Notts, where his parents are buried. The plaque which was originally on the family grave is inscribed: 'To the Glorious Memory of Henry Maudsley DFC, Squadron Leader, RAFVR, killed in the attack on the Möhne dam, 16-17th May, 1943, age 21 years.'[1]

The final casualty was that of Dinghy Young. This time his dinghy did not save him. On his way home from the Eder, having earlier attacked the Möhne, then flown to the Eder in his capacity as deputy leader, with Gibson. Caught by flak west of Amsterdam, his Lancaster crashed in flames into the sea just off the coast. He was buried in Bergen General Cemetery in Holland and his grave has been adopted by Mr and Mrs Dutilh of Bergen. His crew lie with him, having joined as a crew from 57 Squadron. With the exception of David Horsfall, they were all on their first tour. Young was on his third.

Flight engineer was Horsfall, aged twenty-three from Hove, Sussex, who had flown with 106 and 57 Squadrons. The navigator was Sergeant Charles Roberts from Cromer, Norfolk, had only recently joined 57. His grave is tended by Mr Dick Smit of Bergen. The wireless operator, Sergeant Lawrence Nichols, from Westgate, Kent, was thirty-two, while the bomb aimer hailed from Canada, Flying Officer Vincent MacCausland, aged thirty. Front gunner was Sergeant Gordon Yeo, twenty-two, from Barry, South Wales. His body was washed up on 27th May near Wyk-aan-Zee, Holland. Rear gunner was Sergeant Wilfred Ibbotson, aged twenty-eight, a married man from Belton, West Yorkshire.

A total of 56 men failed to return. Of these only three survived as prisoners, two being wounded. The cost of success was in the lives of fifty-three men. Upon his return and de-brief, Gibson wrote in his own hand, fifty-six letters to the families of those who were missing. Any words of hope in them applied in the final analysis to only three lucky ones.

Those who survived the raid, had still to survive the war. Many

[1] Actually the Eder dam.

stayed with 617 Squadron, to attack targets that were tough, difficult or needing, like the dams, special treatment. As the war continued, so it depleted the ranks of those who had flown on the dams raid. More than thirty were killed, mostly in action, a few in accidents. On two days in September 1943, just weeks after their glorious success, nearly a score were lost. All losses were grievous to any squadron, but the most grievous to 617 was the loss of its first Commanding Officer – Guy Gibson.

Gibson flew just one more trip with his squadron, in August 1943, and was then sent to America by Winston Churchill on a goodwill trip, later going on to Canada. They travelled on the *Queen Mary* and included in the party were General Wavell and Orde Wingate. While in Calgary on 8th September, he met the mother of his navigator on the dams raid, Harlo Taerum. She said that it was one of the proudest and happiest moments of her life, and Gibson spent several hours at her home. Eight days later her son with five other members of Gibson's crew were shot down by flak as they made their run-in on the Dortmund-Ems Canal, and killed. She was later to lose another son, Lorner, an air gunner in the RCAF, killed on 3rd February 1945.

Returning to England, Gibson became a staff officer, but managed to fly on a couple of raids. His last came on 19th September 1944, a raid on Rheydt and München Gladbach. He was given special permission to fly a Mosquito (KB267), borrowed from the CO of 627 Squadron at Coningsby, Wing Commander Rupert Oakley DSO, DFC, AFC, DFM. Gibson was Master Bomber for the raid, something he had pioneered over the Möhne.

The weather over the target that night was clear but heavily defended by flak. After he had spotted the target for the bombing force, he was heard to say on the intercom, 'Okay, that's fine, now home.' Three-quarters of an hour later his Mosquito fell near Steenbergen; Gibson and his navigator, Squadron Leader J.B. 'Jim' Warwick DFC, died in the crash that followed. Their remains were buried in the Steenbergen-en-Kruisland, Roman Catholic Cemetery in Holland.

CHAPTER TWELVE

The Damage

In Germany after the attack, a report from the Regierungspräsident Arnsberg, Westphalia, was made to the Minister of Home Affairs. Dated 24th June 1943, the report was translated by Air Ministry dated 22nd July 1947. It contains the text of two German official reports of the raid by the RAF on the Möhne dam, 16th/17th May 1943.

'A number of bombs fell into the reservoir immediately in front of the dam, thereby weakening the structure and probably causing a number of fissures. A further bomb hit the dam itself and aided by the pressure of the contained water mass, made a breach 76 metres wide and 21-23 metres deep.

'An immense floodwave poured through the dam into the Möhne and Ruhr valleys. The main wave carrying about 6,000 cubic metres of water at a height of eight metres, reached a speed of more than six metres per second. At 7.30 am, 1,500-2,000 metres per minute were still flowing through the dam.

'The warden of the dam, Oberförster Wilkening, had a direct line to the exchange at Soest and a tieline to the dam itself and a man on duty there. This line was tested every night according to instructions and was found on the night in question to be in perfect working order. During the raid, however, the terminus at the dam was destroyed by a direct hit and this put the main line to Soest out of order. Oberförster Wilkening finally ran to the station of the Ruhr-Lippe Railway and got through from there, via Koerbecke; this was between 1.10 am and 1.15 am. Between 1.30 and 1.35 am, a second report, warning of an impending flood catastrophe was received.

'The very regrettable loss of human lives could hardly have been avoided in the danger zone because of the size of the breach and the tremendous speed with which the dammed-up water was released.

Had they been informed that there was a possibility of the enemy using bombs of such weight, we should have realised that the existing warning system was inadequate and our technicians would have concentrated on devising a more effective system. Alternatively, the danger area could have been evacuated for the duration of the war.

'Immediately after the catastrophe, I went into the matter with a post office engineer, who made the suggestion that a cable should be embedded in the structure of the dam itself, which would in the event of the structure being affected or agitated, automatically transmit a warning.

'The following damage was caused to overhead and underground electric cables:

100 KV overhead cable – Unna-Nehein – destroyed for 3 km

25 KV overhead cable – Froendenberg-Neheim destroyed for 3 km

10 KV overhead cable – Nierderense-Honninggen – destroyed for 2 km

10 KV overhead cable – Echthausen-Bremen – destroyed for 2 km

10 KV overhead cable – Niederense-Volbringen – destroyed for 1 km

A 25 KV overhead cable, Neheim-Möhne power station is destroyed on a number of places, altogether for about 2 km.

'The camp for workers from the east, which had been erected for the Neheim industries at a cost of nearly a million Marks, was completely destroyed.

'Casualties as on 1st June 1943.

'German dead at Arnsberg, 160 killed and 34 missing; Soest 224 killed and 32 missing; Iserlohn 49 killed, total: 433 dead and 66 missing. Foreigners, 557 dead at Arnsberg and 155 missing, and Iserlohn 6 dead, total: 563 dead and 155 missing. Overall total: 996 dead and 221 missing.

'Up to 22nd May, 1,250 military personnel, 380 men of the technical emergency service, 42 men and 60 women from the Red Cross, and 150 men of the fire service, were used in emergency services.

'It was estimated some 2,000 men would be needed to rebuild the dam etc. The effects of the attack were felt far into the Düsseldorf district.

'The anticipated damage to the water supply system is of the most far-reaching importance. The Möhne dam is the backbone of this system, serving the whole of the right Rhine-Westphalia industrial area, which is inhabited by about 4½ million people. With the loss of the Möhne dam, the available reservoir space had been halved. The result will be that the additional supply of water to the Ruhr will have to be curtailed. The quality of water will be adversely affected and consumption, unless supplies can be got from other sources such as the Lippe, will have to be cut down, even for industrial consumers. It will depend largely on the speed which the repair work can be carried out and also on the weather during the autumn, to what extent industrial – in part war industrial – production will be affected.'

The German Intelligence Branch, prepared a report dated 20th December 1944 which gives an account of the raid on the dams in May 1943 as seen at the time through German eyes. The reader may find interesting the following quotation from the British translation of the report made in October 1945 for the light it throws on the German view of the raids. Being based on incomplete or erroneous information the report is confused and inaccurate and among the obvious errors which will be found commented upon at the end of the chapter it will be noted that the sequence in which the aircraft are stated to have attacked the Möhne Dam does not tally with the facts of British records and the cool observations of RAF aircrews.

'Explosion No 3 produced a considerably smaller movement than explosion 1, 2 and 4. Explosions 5 and 6 were caused, according to the time and allowing for the distance, by the mines dropped on the Eder dam. It is believed that the movement noted at 337hrs was due to an explosion approximately 180km west of Göttingen. Taking the times into account only the bombs given under 1-4 need to be considered in relation to the destruction of the Möhne dam. The first mine released against the Möhne dam must, according to observation by numerous people, have exploded in front of the torpedo netting and so have torn the net barrage. The torpedo netting was fixed at about 25 metres from the water side of the wall to floating buoys 350 metres apart. It consisted of two nets 6 metres apart, stretching down to a depth of 15 metres below the edge of the overflow. The upper rope of the nets stretching right across the water surface was secured at both shore ends. In the middle on the water side of the net a heavy flat barrel was anchored. A rope was carried from the net over a pulley fixed to this barrel to a cement weight which was designed to keep the net at a distance from the wall in the centre by the pressure on the rope, even if the water level varied. At both sluice towers this distance was ensured by floating wooden fenders 30 metres in length. The spot at which the mine struck the net cannot be definitely established but we may assume that the point of impact was near the left hand sluice tower (looking down the valley) at a point where the floating fender is badly damaged. The flat buoy is entirely undamaged, also the anchor rope. The three anchor weights have only been shifted a few metres down the valley.

'There is no doubt about the point of impact of the second mine which exploded near the left bank about 80-100 metres from the wall. The swell set up by this mine caused the loose earth on the left bank to collapse over a considerable distance. When the dam was emptied parts of this mine were found. The occupants of the left sluice tower heard a heavy cascade of water at the left-hand end of the dam wall which led them to believe that the dam wall had been breached to the left of their tower. Actually the left-hand section of the wall is undamaged. The flow of water clearly only poured over the top of the wall.

'The aircraft which released this mine appears to have approached too close with the result that the mine approached the wall at an angle and exploded at a relatively great distance from the wall. The aircraft which dropped the third bomb must also have approached the exact centre of the wall but clearly released the bomb too late. It went clean over the top of the wall and fell into an overflow pond situated below the dam wall. This caused the destruction of a power plant built right across the valley about 40 metres from the open side of the wall and it blew up with a large sheet of flame. Simultaneously the electric current failed over the whole valley area. The water in the overflow pond was only a few metres deep. It is therefore comprehensible that the effect of this mine produced only a week record in Göttingen since the most of the explosion was in the air and caused considerable house damage in the vicinity of the dam wall. On the other hand the effects of both of the first two mines, which exploded deep under water, on nearby houses was relatively slight. At the Seehof Inn, situated next to the right hand end of the dam wall, only a few windows were broken. The deep underwater blanketing effect directed the force of these charges to a large part of the rocky bottom and caused a violent extension of the impact of the explosion as far as Göttingen. The blast of the third mine which dropped on to the overflow pond caused the roof of the left hand sluice tower to collapse and blew the gun posted there from the flat roof making it impossible to serve the gun. Thereupon the crew left the tower in order to assist the gun crew of the right hand sluice tower in carrying ammunition. The gun crew were still able to proceed by way of the undamaged roadway of the centre portion of the wall to the right hand tower. Soon after they had reached it they observed a further heavy aircraft approaching which flew very low and released a fourth mine a few metres from the centre of the wall. After a dull explosion which did not appear to the crew to be particularly heavy the wall between the two towers collapsed and water poured into the valley.

'These facts prove a total of 3 mines was dropped on the water side of the Möhne dam. I consider particularly important the fact that these three mines were not all dropped at the point where the breach occurred, but exploded at relatively widely separated points. This would appear to indicate that the last mine which exploded close to the centre of the wall was able to cause the breach without previous serious damage having been sustained by the wall as a whole. This appears to be borne out by the horizontal cracks observed in the dam wall when the water had been drained off. These cracks are confined to the area between the two sluice towers and are approximately symmetrical with the breach itself.

'We may assume that the first mine dropped on the Eder wall, which clearly exploded at a considerable distance from the wall, caused horizontal cracks on the whole centre section of the wall, which was in any case thinner, right down to the centre section of the wall, whilst the second mine which clearly exploded nearer to the foot of the wall caused a breach at a short distance from the centre of the wall.

'We can now describe the bombing attacks on the Sorpe dam which were carried out the same night at about 0045hrs by a single aircraft from an approximate height of 20 metres. The dam was not protected by AA or balloon barrage or nets.

'The aircraft flew several times across the dam all in the direction of the earth dam and did not release a heavy mine till the tenth run. This formed a bomb crater on the water side of the earth wall close to the water line covered only by 3 metres of water with the result that, since it blanketed the explosion only to a minor degree, this explosion took an upward direction and threw up a column of water to a height of 150-200 metres. The downward effect of the explosion was damped by the material of the rubble dam and this explains why the effects of the explosion were not felt in Göttingen. As the first bomb failed to breach the earth dam a second attack was made on the Sorpe dam shortly after 3 o'clock which was clearly carried on to the Bever dam. Only one mine was dropped and fell about 800 metres from the dam wall in the middle of the reservoir and did no damage.'

At the same time as the Sorpe dam was attacked, the Ennepe dam was attacked by a single aircraft. It dropped its bomb which fell about 800 metres from the dam wall in the middle of the reservoir. The damage caused to the Möhne dam as as follows: An approximately rectangular wedge 76 metres broad and of average depth of 20-22 metres was broken away. The wall approximately 17 metres thick at the back of the breach. Approximately 12,500 cubic metres of masonry was washed away at the breach.

As the Möhne dam was built in a steep curve, the section which was destroyed exerted a strong downward pressure on adjoining wall sections and this caused further considerable damage. When these sections were demolished, deep fissures were detected which had undermined the stone work to such an extent as to make its removal imperative. An additional 6,800 cubic metres of stone work were removed with a result that 1,950 cubic metres of wall needed replacing.

The final toll of losses was: 1,294 people drowned, 6,500 cattle and pigs lost, 125 factories destroyed or damaged, 3,000 homes ruined, 25 bridges submerged and 21 badly damaged.

On 18th May 1980 the water level of the Möhne reservoir was as high as it had been in May 1943. The breached section of the dam is still easily distinguishable. The repair work took the force of workers until 23rd September 1943, in time to catch the autumn

Möhne Dam being rebuilt after the raid

Möhne Dam after the raid

rains. The Germans then thought the British would bomb the dams again, or especially while they were under repair. Normal bombing only would have been required, which would have produced a cave in, blasted masonry and set fire to the wooden scaffolding used for the rebuilding. It had been Wallis's idea that once destroyed, the dams should remain out of action by bombing them from time to time before repair work could be completed, but they never were.

The first troops to reach the Möhne dam were the 8th US Tank Division who arrived on 7th April 1945. When they arrived all the defence precautions were still in place, torpedo nets and steel masts with contact mines attached. The British arrived in June 1945. That year the Möhne was reinforced with concrete and faced with masonry similar to that used in the original construction. In 1946 a visit by RAF personnel from Dortmund airfield, No 2001 HQ Staff, Admin. Wing, found the wall on the side of the dam still down, in its place a wooden rail.

There is nothing on the dam to remind visitors today of the breach. A plaque which once resided in one of the towers, now lies in Canberra Museum in Australia, the marks where the plaque once hung can be clearly seen. A new power house was built in 1953, on the south side, rather than set directly under the dam as the original one had stood.

The area of the dam is now a tourist site for camping and weekend visitors. Here one can take a boatride on the reservoir or dine at a very nice restaurant built to overlook the dam.

The Eder is much higher than the other dams, and one can in fact drive a car over it. The area of the breach can be worked out by the outsets set into the wall of the dam. On the left there are two and on the right, seven at least. Two appear to be missing, not being replaced when the dam was rebuilt. Now the Sorpe is lined by hedgerows and one can walk the complete length of it. The water, like the Möhne, was up to the level of 1943.

Of the many inaccuracies in the above probably the most significant are:

five bouncing bombs were launched at the Möhne Dam not four as claimed:

four exploded on the water side of the barrage not three as claimed:

full breach of the Möhne Dam occurred only after the fifth weapon was released not the fourth:

three weapons were launched at the Eder Dam not two:

the Eder broke after the third weapon exploded not the second:

the bouncing trajectory of the bomb launched by Gibson's crew was not deflected by the torpedo boom because it was seen to explode accurately and centrally by his own and other RAF aircrew; any one of the five weapons dropped may have cracked the torpedo boom if indeed that was the cause of the boom cracking and not being carried down by the water outpour:

pieces of bomb casing were in fact blasted over a wide area and where the bits were picked-up cannot be related to the point of detonation:

the second aircraft to attack the Möhne Dam was shot down not the third, otherwise Hopgood would be alive today and not Martin:

the claim that the third weapon exploded near the lefthand or south side of the dam wall is negated by AA Gunner Schutte's own report which says '… then came a heavy explosion and a great water spout, again the lake quaked and waves engulfed the wall'. According to his own evidence he had *by then* run from the south tower to the north tower because Hopgood's bomb exploding over the dam wall had blasted the gun platform off the south tower, which means Schutte witnessed the third bomb explode between the two towers but nearer the south than the north:

no aircraft attacked at a shallow angle – that is pure surmise:

the collapse of any part of the earth banks around the lake is valueless evidence as the banks folded in at many places at both the Möhne and Eder Dams as the waters surged or drained out:

photo/reconnaissance of the Ennepe (F/O Scott EN141) on 18th May showed the Ennepe had been damaged.

CHAPTER THIRTEEN

The Legend is Set

One thing that comes out in any war is comradeship, which turns to lasting friendships. In the case of 617 Squadron, these two factors are perhaps just that little bit more evident than in other squadrons.

Since the early 1950's they have been meeting regularly. The publication of Paul Brickhill's book *The Dam Busters* in 1951, brought them together for the first occasion. In 1955, the now famous film based on the book was released, and a World Première held over two days at the Empire Theatre, Leicester Square in London. It was held on 16th and 17th May, the twelfth anniversary of the raid on the dams. It had to be held over two days as there were so many people to be invited. Among those attending on the 16th were:

R.E. Grayston, H.S. Hobday DFC, E.C. Johnson DFC, W.C. Townsend CGM, DFM, D.E. Webb DFM, H.B. Martin DSO, DFC, I. Whittaker DFC, B.T. Foxlee DFM, B. Goodale DFC, J. Buckley DFC, E.E. Appleby, W. Howarth DFM, G. Rice DFC, H.J. Hewstone; all original members of the squadron, and all were presented to HRH Princess Margaret. Also present were Mrs Jack Hyman, formerly Mrs Eve Gibson, and Mr Alexander Gibson, Guy's father.

The next evening the following were presented to their Royal Highnesses, the Duke and Duchess of Gloucester:

J.C. McCarthy DSO, DFC, D.R. Walker DFC, D.A. McLean DFM, K.A. Brown CGM, P.E. Pigeon DFC, R. Wilkinson DFM, F. Tees, H.B. Feneron, D.P. Heal DFM, A.F. Burcher DFM, L.J. Sumpter DFC, DFM, C.E. Franklin DFM. Both Tony Burcher and Freddie Tees had been prisoners after the raid.

Also present were Mr J. Whillis, brother of the late Pilot Officer S.L. Whillis who had been in Barlow's crew; Mr A. Yeo, father of the late Sergeant C.A. Yeo, of Young's crew; Mr S. Pulford, brother of the late Sergeant John Pulford DFM of Gibson's crew.

The film, made in black and white, was directed by Michael Anderson, and starred Richard Todd as Guy Gibson, and Michael Redgrave as Dr Barnes Wallis, a part he played to perfection. Both Richard Todd and Michael Redgrave attended the première. Todd recently said that he felt very privileged to have portrayed Gibson, and to be associated with his famous squadron.

Two years had been spent in preparing and researching the film. Richard Todd spent a good deal of time talking to people who had known Gibson, such as Barnes Wallis, Gibson's wife and father, Mick Martin and even a former schoolmaster.

Wing Commander Wally Dunn OBE was the technical advisor for the film. In July 1954 he was invited by Michael Anderson to watch the filming of the reconstruction of the attack on the Möhne dam. They went into three huge sheds, called stages, and had to go through two steel doors, which were locked shut before the filming began. Part of the floor had been removed, disclosing a sheet of water in the shape of the Möhne lake, 150 feet wide, 300 feet long. It was lined with grass and imitation fir trees, gradually rising up to rocky hills on each side. In the middle was a perfect model of the dam, just like the small model Wally Dunn had seen at Grantham eleven years earlier. Alongside the set was a two feet track on which a high speed camera trolley had been mounted, with a cine camera with an aircraft-type seat behind it, both being placed on a long steel trellis swivelling arm. The cameraman was strapped into the seat and the trolley moved back ready to start its run. There was a shout of 'Silence – attack on the Möhne, scene 227, take one, roll!'

The trolley raced down the track with the camera shooting a side view of the approach to the dam and as it came level with it, a high column of water rose into the air to give the impression of a bomb exploding. Washing powder encouraged the water to foam. The trolley carried on, filming the water spilling over the dam wall, then beyond until the trolley stopped at the end of the set. The actual flying scenes with Avro Lancasters were shot at Derwent lake reservoir, near Sheffield, except that the underbelly bomb was not accurate. The bomb was at that time still on the secret list and the film makers were not allowed to portray what the actual bomb had looked like. They interpreted the 'bouncing bomb' as being more

Bomb away

A practice bomb,
Reculver Bay

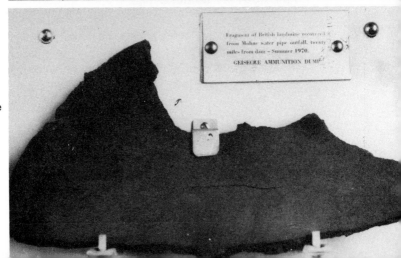

ce of a bomb recovered
y miles from the Möhne
in 1970

ball-like, and this is exactly how the film bomb looked like.

When the film had been made, Wally Dunn dubbed the morse onto the sound track for the operations room scene. He was put into a soundproof cabinet and through a window he could see a film of the action taking place in the operations room at Grantham. He watched himself on the dais in the ops room and at the appropriate moment, transmitted the morse code, on a service morse key and valve oscillator, actually dubbing the film with the morse of each aircraft. A similar sequence was undertaken for the attack on the Eder dam film.

As a run-up to the film première, a reception was given on 14th May, for the current 617 Squadron, by the RAF Association, British Picture, and the Pathfinder Association, held in the Louis XV room, at the Criterion Restaurant in London.

To mark the 25th Anniversary of the Dams Operation, a Thanksgiving service was held at St Clement Danes Church, on 19th May 1968. It had been preceded on the 18th May by a reception which started with a showing of the film, at the Warner Theatre, Leicester Square, put on by the Associated British Picture Corporation and Warner Pathé. Nearly all the survivors were there, also Barnes Wallis. After the film there was a cocktail party, attended by Richard Todd, Sir Arthur Harris, Sir Ralph Cochrane, and Charles Whitworth, former station commander at Scampton.

The 'Thanksgiving' service was presided over by the Reverend (now Canon) J.E.R. Williams DFC, a former pilot with 617. Richard Todd delivered a passage from the Funeral Oration of Pericles, and the lessons were read by Mick Martin and Leonard Cheshire VC, who commanded 617 later in the war. A reception afterwards ended the weekend on a note of pleasure, nostalgia and some sadness as they remembered the men who could not make the reunion; they had given their today for all our tomorrows.

Four years later, in 1972, a reunion was organised in Canada. It was a week of parties, ending with a party in Toronto, to end all parties! About forty Canadian ex-members of the squadron attended, and of those, eleven were original members, including Ken Brown, Danny Walker, Don McLean and Dave Rodger.

In September 1976 the reunion took place in Holland. The party left Victoria Station at 8.30 am on the morning of 18th September, crossing the Channel from Dover to Amsterdam, arriving that night at the Amsterdam Hilton Hotel. The next day they made a visit to the military cemetery at Arnhem, and then went on to Bergen Op Zoom Cemetery, where Lewis Burpee and his crew are buried, all lost on the dams raid. In the afternoon came a trip to Steenbergen to visit the graves of Guy Gibson and his navigator Jim Warwick, where a short service was conducted by Canon John Williams. Garlands were laid by the Council of Steenbergen, the RAF Association, the Dutch Resistance and former members of 617. This was followed by the playing of the National Anthem and a flypast by the only Lancaster aircraft still flying, a Vulcan bomber and four Harvard aeroplanes.

The following year revolved around the 90th birthday of Barnes Wallis, by then Sir Barnes. Wallis had been made a CBE in 1943 following the dams raid, and then on 8th June 1968 was made a Knight Batchelor.

A splendid dinner was given for him by 617's wartime members, held at the Thatchers Hotel, East Horsley, in Surrey. Included in the menu was 'Wellington Cocktail', 'Tournedos Barnes Wallis', and 'Airship Meringue', all specially prepared by the Chef, David Taylor. Guests including Sir Arthur Harris, Sir Ralph Cochrane, Gibson's widow Mrs Eve Hyman, joined Danny Walker, Don McLean, Dave Rodger, Lance Howard, Bill Townsend, Johnny Johnson, Hobby Hobday, Bob Kellow and Jack Buckley.

Sir Barnes died on 30th October 1979 at the age of 92. He had predicted the year in which he would die some years earlier. He had gone into hospital just for a rest, and had been there about eight days when he died peacefully in his sleep.

Also in 1977, the 34th Anniversary of the raid, a flypast was arranged over the Derwent Reservoir by the flying Lancaster, watched by more than 5,000 interested spectators. On the dam itself, stood six of the original members of 617, Bill Townsend, Ivan Whittaker, Bill Howarth, George Chalmers and Jack Buckley. Among the spectators was a local farmer, Ray Ollerenshaw. He recalled seeing the original practice runs of the lake in 1943. He remembered what a tremendous sight it had been, as the Lancaster skimmed over the water, much lower than the run made by the Lancaster they had just watched. In 1943 the Lancasters had shot out below the level of the two towers after dropping their practice bombs.

Another man to recall an incident concerning 1943, during the 1977 celebrations, was John F. Grime DFC. He had been posted to 207 Squadron on 16th May 1943 at RAF Langar, Nottingham. That night, 207 were on stand-down, but being a new boy he and his crew were down for a high level bombing practice in the morning, and a Bullseye' exercise that night – the night of the raid. This latter trip turned out to be a four hour sortie to the French coast, then a flight up and down the Channel. Little did they know they were in some way a form of help in forming a small part of the diversions for the 617 raiders. The next day, John and the others in his squadron, saw photos of the breached dams.

John Grime had in fact planned a 21st birthday party for that night and had been a little upset about having to fly on the exercise, but later felt that it had been worthwhile. In any event they all enjoyed the postponed party the next night.

In September 1979 a plaque in memory of Wing Commander Guy Gibson was presented to the Royal Air Force Museum at Hendon. It came from the wartime Dutch Resistance Fighters and is on permanent location near the experimental equipment used and designed by its inventor, Barnes Wallis.

To this day, RAF Scampton is very much as it appeared in 1943. Just a twenty minute bus journey from the centre of Lincoln, with its towering Cathedral set above the city – a welcome landmark for many a wartime flyer coming home – and the homing point for many wartime crews. The hangars house Vulcans where once they housed Lancasters. In No 2 Hangar, the old home of 617, the present squadron still reside. Nearby is the operations room and their small, but interesting museum. On its walls are many photographs of the dams and they even have one of Guy Gibson's service caps. The present squadron are very proud of their wartime squadron and although looking after the museum is a secondary job to their flying duties, they do a magnificent job in answering many enquiries about the dams from a still very interested public.

Barnes Wallis – 1978

o R: Ernie Twells, Bill
wnsend, Ivan Whittaker,
Howarth, George
almers, Jack Buckley,
ampton, May 1977

: Douglas Webb, Bill
send, Mick Martin,
Feneron, Jack Buckley,
y Johnson, Toby Foxlee,
wn, Bill Howarth,

All the later Commanding Officers of 617 Squadron, use the office used by Guy Gibson in 1943. Looking through the window one sees the panorama of the airfield where once Lancasters stood and where now Avro Vulcans stand. To the right, a small but very famous grave of a dog named Nigger.

The latest reunion took place in April 1980 in Australia. While there, the 617 contingent led the Anzac parade in Adelaide, and laid a commemorative tree for the squadron at Adelaide Airport. On Sunday the 27th, a plaque was also unveiled at the airport, its inscription reads: 'Erected in memory of gallant comrades Royal Australian Air Force, who served in 617 RAF 'Dam Buster' Squadron, 1943-1945. Lest we Forget.' It was unveiled by Ross Standford DFC, President of the 617 Squadron Association in Australia. Wreaths were laid by Air Commodore H.V. Gavin OBE, DFC, who had been among the force of 617 which attacked the *Lützow* in 1945, Group Captain J.B. Tait DSO, DFC, who had commanded the squadron in 1944. Squadron Leader D.A. Bell DFC, Squadron Leader Les Munro DSO, DFC, and Captain H. Knilans DSO, DFC, USAF, an American who had flown with 617 in 1944, having joined the RCAF before America came into the war.

Following the Australian trip, a week-end reunion was arranged at Rolls-Royce in Derby, England for 16/17th May 1980. On the 16th they had a buffet dinner at the Assembly Rooms in Derby, followed the next day by a tour and cavalcade through Derby in vintage cars. At 4 o'clock in the afternoon, a helicopter flew the 50-mile journey to the Derwent dam to drop a wreath of poppies in memory of the men lost on the dams attack. Among the original crew members who flew in the helicopter, were: Basil Feneron, Bill Townsend, Mick Martin, Bill Howarth, Jim Clay, Geoff Rice, Roby Foxlee, and Johnny Johnson. During the weekend a painting of Sir Harold (Mick) Martin's Lancaster, P-Popsie, was presented to the CO of 617 Squadron, Wing Commander John Herbertson. Also that year a bust of Guy Gibson was presented to the 617 Association by a Dutch couple.

The legend of the Dambusters never seems to die. Hardly a year goes by without hearing, seeing or reading something to do with their past history. In January 1981, a gold cigarette case which had belonged to Guy Gibson was sold at auction, inscribed: 'ED932-G G.P.G., May 17th, 1943. "Nigger" "Dinghy" ' – both code words for breaching of the Möhne and Eder dams. It was purchased by Mr Peter Skinner, himself a former RAF pilot. The case had been presented to Gibson after the dams raid, by Vickers Armstrong.

Even a flower has been named in memory of the Dambusters. The President of the Birmingham Dahlia Society, George Brookes, named a dahlia after them, another after Guy Gibson, and a third after Barnes Wallis. George Brookes himself was involved in the horrific business of Bomb Disposal in many parts of England in the war. In this respect he has something in common with 617, both being involved with bombs.

All over the world, the name Dambusters, has become a legend. It has no age barrier, young and old alike are captivated by this one unique operation in the history of the Royal Air Force in WW2. As far as Bomber Command are concerned, it is parallel with the interest in the Battle of Britain, as a short but vital period in the history of Fighter Command.

Now thirty-nine years later, the men who remain of those who flew out that May night in 1943, are a little thinner and greyer on top, and perhaps a little larger in the waistline, but are still very much the men who made that historical mission. In a short six-week period that these men were thrown together, came a spirit and comradeship which all these years later is still there, as strong as ever. At their gatherings and reunions, they talk over old times when they flew together, and although the years have rolled by, all those years seem to disappear and they are back to those days in 1943, talking of Gibbo, Dinghy, Hoppy and all their other old comrades.

All the time they were building up to that one day, a day they set the world alight by punching a hole right at the enemy's heart, and were within a whisker of tearing it out. A legend yes, when ever World War Two is thought of, talked of, or read about, the name 'The Dambusters' and their story will always be in the forefront.

The Aircraft

ED864 'B' – Tested 16th April; To 617 22nd April; Flown by Bill Astell. Lost on Dams Raid; Total hours flown, 22.

ED865 'S' – Delivered to 617 22nd April; Flown by Lewis Burpee. Lost on Dams Raid; Total hours flown 17.

ED886 'O' – Delivered to 617 23rd April; Bill Townsend's aircraft on the Raid. Subsequently lost 11th December 1943, having flown a total of 138 hours.

ED887 'A' – Delivered to 617 22nd April; Flown by Dinghy Young on the Raid and was lost.

ED906 'J' – Delivered to 617 23rd April; Flown on the Raid by David Maltby. Damaged 20th March 1945, returned to 617 18th May 1945; To No 46 MU 24th June, back on 617 27th August 1946. Scrapped 29th July 1947.

ED909 'P' – Delivered to 617 23rd April; Flown on the Raid by Mick Martin and was damaged. Returned to 617 after repair 12th June. Damaged 30th October 1943 and back on 617 20th November. Became 6242-M, to No 46 MU 29th January 1945, and back again to 617 13th August 1946. Scrapped 29 July 1947.

ED910 'C' – Delivered to 617 28th April; Flown on the Raid and lost, Warner Ottley's aircraft; Had flown 20 hours.

ED912 'N' – Delivered to 617 3rd May; Flown on the Raid by Les Knight. To No 46 MU February 1945; Scrapped 26th September 1946.

ED918 'F' – Delivered to 617 30th April; Flown on the Raid by Ken Brown, and damaged; Returned to 617 29th May; Crashed and burnt out 21st January 1944.

ED921 'W' – Delivered to 617 30th April; Flown on the Raid by Les Munro, and damaged. To No 45 MU 29th January 1945; To A.V. Roe 24th May 1945. Scrapped 26th May 1946.

ED923 'T' – Delivered to 617 2nd May; Flown on the Raid by Joe McCarthy. Subsequently lost 8th September 1943 after 75 flying hours.

ED924 'Y' – Delivered to 617 30th April; Flown on the Dams Raid by Cyril Anderson. Damaged 2nd July 1944, to No 46 MU 1st February, 1945. Scrapped 23rd September 1946.

ED925 'M' – Delivered to 617 30th April; Flown by Hoppy Hopgood on the Raid and shot down. Had flown a total of 17 hours.

ED927 'E' – Delivered to 617 3rd May; Flown by Robert Barlow on the Raid and lost. Total flying hours 20.

ED929 'L' – Delivered to 617 30th April; Flown on the Raid by Dave Shannon; To No 46 MU 29th July 1945. Scrapped 7th October 1946.

ED932 'G' – Delivered to 617 30th April; Flown on the Raid by Guy Gibson; To 467 Squadron 7th February 1945; To 61 Squadron 27th August 1946. Scrapped 29th July 1947.

ED934 'K' – Delivered to 617 3rd May; Flown on the Raid by Vernon Byers, who failed to return; Total hours flown 13.

ED936 'H' – Delivered to No 39 MU 4th May; To 617 12th May; Flown on the Raid by Geoff Rice, and damaged; Returned to 617 17 July 1943.

ED937 'Z' – Delivered to No 37 MU 6th May; To 617 14th May; Flown on the Raid by Henry Maudsley who failed to return. Had flown 7 hours.

APPENDIX II

The Awards
Gazetted 28th May 1943

Victoria Cross

Wing Commander G.P. Gibson DSO, DFC

Distinguished Service Order

Flight Lieutenant L.G. Knight
Flight Lieutenant J.C. McCarthy DFC
Squadron Leader D.J.H. Maltby DFC
Flight Lieutenant H.B. Martin DFC
Pilot Officer D.J. Shannon DFC

Conspicuous Gallantry Medal

Flight Sergeant K.W. Brown
Flight Sergeant W.C. Townsend DFM

Bar to Distinguished Flying Cross

Flying Officer B. Goodale DFC
Flight Lieutenant R.C. Hay DFC
Flight Lieutenant R.E.G. Hutchinson DFC
Flight Lieutenant J.F. Leggo DFC
Flying Officer D.R. Walker DFC

Distinguished Flying Cross

Flight Lieutenant J. Buckley
Flight Lieutenant L. Chambers
Pilot Officer G.S. Deering

Pilot Officer J. Fort
Flight Lieutenant H.S. Hobday
Flight Lieutenant C.L. Howard
Flight Lieutenant E.C. Johnson
Flight Lieutenant R.A.D. Trevor-Roper DFM
Pilot Officer F.M. Spafford DFM
Pilot Officer T.H. Taerum

Bar to Distinguished Flying Medal

Flight Sergeant C.E. Franklin DFM

Distinguished Flying Medal

Flight Sergeant G.S. Chalmers
Flight Sergeant D.P. Heal
Sergeant G.L. Johnson
Flight Sergeant D.A. McLean
Flight Sergeant V. Nicholson
Sergeant S. Oancia
Flight Sergeant J. Pulford
Flight Sergeant T.Dm Simpson
Flight Sergeant L.J. Sumpter
Sergeant D.E. Webb
Flight Sergeant R. Wilkinson

The Men

Wing Commander Guy Penrose Gibson: Pilot and Squadron Commander
b. 12th Aug 1918, Simla, India.
Killed in Action 19 September 1944

SERVICE	PROMOTIONS	AWARDS
Joined RAF 1936		
CFS Yatesbury 16 Nov 1936		
No. 24 Depot 31 Jan 1937	Acting Pilot Officer	
No. 6 FTS 6 Feb 1937	16 Nov 1936	
83 Sqdn 4 Sep 1937	Flg Off 16 June 1939*	DFC Gaz
No. 14 OTU 26 Sep 1940	Flt Lt 3 Sep 1940	9 Jul 1940
No. 16 OTU 10 Oct 1940		
29 Sqdn 13 Nov 1940	A/Sdn Ldr 29 Jun 1981	bar DFC
No. 51 OTU 23 Dec 1941	T/Sdn Ldr 1 Dec 1941	16 Sep 1941
No. 51 Gp HQ 23 Mar 1942	S/L 13 Apr 1942	
106 Sqdn 13 Apl 1942	A/Wg Cdr 13 Apr 1942	DSO Gaz
		20 Nov 1942
5 Gp HQ	15 Mar 1943	bar DSO
		2 Apl 1943
617 Sqn 24 Mar 1943		VC Gaz
Special Duties Aug 1943		28 May 1943
Air Ministry 3 Jan 1944		
28 Gp Course 13 Mar 1944		
No 54 Base HQ 12 June 1944		

Sergeant John Pulford (652403) Flight Engineer
b Hull.   Motor Mechanic.
Killed in Flying Accident 13 February 1944.

Joined RAF 1939		
No. 3 Depot 9 Aug 1939		
No. 3 Wing 22 Sep 1939		
10 Sqdn 1 Apl 1940		
RAF Kenley 25 Sep 1940		
No. 1 STT 14 Jan 1942		
No. 20 RC 3 Jun 1942		
No. 4 STT 9 Sep 1942		
207 Sqdn 12 Oct 1942		
No. 1660 CU 20 Oct 1942		
97 Sqdn 25 Nov 1942	Flew 10 operations	
617 Sqdn 4 Apl 1942		DFM Gaz
	Temp F/Sgt 8 Oct 1943	28 May 1943

* All promotion dates in this list are the effective date, so these promotions will have
been actually made after the given date.

Pilot Officer Torger Harlo Taerum RCAF (J/16688) Navigator:
b Milo Canada, 22 May 1920.
Killed in Action 16 September 1943.

SERVICE	PROMOTIONS	AWARDS
Joined RCAF 1941		
ITS 27 Jan 1941		
BLG 7 Jun 1941		
AOS 27 Apl 1941		
ANS 7 July 1941		
50 Sqdn Mar 1942	28 Operations	
No. 1654 CU 24 Feb 1943	F/Sgt Apl 1943	
617 Sqdn 3 Apl 1943	Plt Off 24 Jun 1943	DFC Gaz
	A/Flt Lt Sep 1943	28 May 1943

Flight Lieutenant Robert Edward George Hutchinson (120954) Wireless Operator:
b. Liverpool, 1918.
Killed in Action 16 September 1943.

RAFVR 1940		
No. 3 RC 6 Jan 1940	(No. 977611)	
RAF Upwood 23 Feb 1940		
No. 2 E & WS 24 Jun 1940		
RAF Invergordon 22 Feb 1941		
No. 2 Sig Sch 3 Jul 1941		
No. 5 B & GS 8 Aug 1941		
No. 25 OTU 31 Aug 1941	Plt Off 20 Apr 1942	
106 Sqdn 8 Dec 1941	Fl Off 20 Oct 1942	
No 1654 CU 18 Mar 1943	Flew 33 operations	DFC Gaz
		12 Feb 1943 (28 ops)
617 Sqdn Mar 1943	A/Flt Lt 20 Apl 1943	Bar DFC
		28 May 1943

Pilot Officer Frederick Michael Spafford RAAF (A/407380) Bomb aimer:
b 16 Jun 1918, Wayville, Adelaide, Australia.   Fitter.
Killed in Action 16 September 1943.

Joined RAAF 1940		
Trainee 14 Sep 1940	Sgt 6 May 1941	
To England 27 May 1941		
83 Sqdn	F/Sgt 6 Nov 1941	
455 Sqdn		
50 Sqdn 24 Apl 1942	Flew 33 operations	DFM Gaz
No. 1660 CU 17 Jan 1943	Plt Off 15 Jan 1943	20 Oct 1942 (15 ops)
617 Sqdn Mar 1943	Fl Off 15 Jul 1943	DFC Gaz
		28 May 1943

Flight Sergeant George Andrew Deering RCAF (J/17245) Front Gunner:
b Ireland.   Shoemaker.
Killed in action 16 September 1943.

Joined RCAF 1940		
1941	Sgt WOP Feb 1941	
Bomber Sqdn	Flew 35 operations	
617 Sqdn 29 March 1943	Plt Off 18 May 1943	DFC Gaz 28 May 1943

Flight Lieutenant Richard Algernon Dacre Trevor-Roper (47354) Rear Gunner:
b 19 May 1915, Shanklin, IoW.
Killed in Action 31 March 1944.

SERVICE	PROMOTIONS	AWARDS
2/Lt RA Aug 1935		
Joined RAFVR 1939		
RAF Cardington 16 Aug 1939		
No. 2 ELWS 14 Sep 1939		
RAF Yatesbury 16 Apl 1940		
No. 12 Obs Sch 7 Jun 1940		
No 1 Sigs Dep 23 Jun 1940		
No. 4 B&G Sch 18 Jul 1940		
No. 16 OTU 11 Aug 1940		
RAF Finningley 13 Oct 1940	Plt Off 24 Oct 1941	
50 Sqdn 27 Oct 1940		DFM Gaz
55 Sqdn 16 Jun 1942		23 Dec 1942 (24 ops)
No. 61 CO 23 Jul 1942		
No. 1660 CU 27 Oct 1942	Fg Off 1 Oct 1942	
50 Sqdn 14 Nov 1942	Total ops flown – 51	
617 Sqdn 25 Mar 1943	F/Lt 24 Oct 1943	DFC Gaz
CGS, 25 Gp 24 Aug 1943		28 May 1943
RAF Upwood 19 Feb 1944	Pathfinder Training	
97 Sqdn 5 Mar 1944		

Flight Lieutenant John Vere Hopgood (61281) Pilot:
b 29 Aug 1921, London.
Killed in Action 17 May 1943.

Joined RAF 1940		
No. 5 ITW 12 Aug 1940		
No. 50 Pool 21 Sep 1940		
Cranwell 30 Oct 1940	P/O 16 Feb 1941	
No. 2 Nav Sch 22 Feb 1941		
No. 14 OTU 26 Apl 1941		
50 Sqdn 10 Jul 1941		
No. 25 OTU 24 Oct 1941		
106 Sqdn 25 Feb 1942	F/O 16 Feb 1942	DFC Gaz
		27 Oct 1942 (32 ops)
No. 1660 4 Nov 1942	F/Lt 16 Feb 1943	Bar DFC
617 Sqdn 31 Mar 1943		12 Jan 1943 (45 ops)

Sergeant Charles Brennan (942037) Flight Engineer:
Killed in Action 17 May 1943.
Joined RAF 1939
No. 2 RC 29 Nov 1939
No. 8 RT Pool 12 Dec 1939
RAF Locking 19 Jan 1940
No. 7 STT 21 Jun 1940
RAF Warmwell 19 Sep 1941
RAF Ibsley 5 Nov 1941

Sergeant Charles Brennan (942037) Flight Engineer: (continued)

SERVICE	PROMOTIONS	AWARDS
66 Sqdn 12 June 1942		
No. 4 STT 30 Jun 1942		
106 Sqdn 30 Jun 1942		
No. 207 CF Oct 1942		
No. 1660 CU 20 Oct 1942		
617 Sqdn Mar 1943		

Flying Officer Kenneth Earnshaw RCAF (J/10891) Navigator:
b Alberta, Canada.   Teacher.
Killed in Action 17 May 1943.

50 Sqdn 9 Nov 1942	Flew in crew
with Trevor-Roper	
617 Sqdn Mar 1943	

Sergeant John William Minchin (118097) Wireless Operator:
b 1916
Killed in Action 17 May 1943.
Joined RAF 1940
No. 2 RC 20 Jul 1940
No. 9 RC 22 July 1940
RAF Cardington 10 Aug 1940
No. 18 OTU 30 Sep 1940
No. 10 Sig RC 28 Oct 1940
No. 2 W & SS 15 Feb 1941
No. 1 A&G Sch 31 May 1941
No. 16 OTU 12 July 1941
49 Sqdn 29 Oct 1941
No. 26 OTU 31 Aug 1942
RAF Scampton 9 Apl 1943
617 Sqdn Apl 1943

Pilot Officer J. W. Fraser RCAF (J/17696) Bomb aimer:
Became prisoner of war 17 May 1943 and sent to Stalag Luft III.
After the war returned to Canada and became a forest ranger, but has since died.

50 Sqdn		
617 Sqdn Mar 1943	Plt Off 14 Mar 1943	

Pilot Officer George Henry Ford Goodwin Gregory (141285) Front Gunner:
from Glasgow.   Printer.
Killed in Action 16 May 1943.

Joined RAF 1939		
1939	(No. 755905)	
44 Sqdn 1941-1942		DFM Gaz
No. 16 OTU 8 Oct 1942	Plt Off 8 Oct 1942	22 Aug 1941 (9 ops)
617 Sqdn 9 Apl 1943		

Flying Officer Anthony Fisher Burcher RAAF (A/403182) Rear Gunner:
b Sydney, Australia, 1922.
Prisoner of war 17 May 1943 and sent to Stalag Luft III.
Repatriated to England 14 May 1945, returned to Australia 23 January 1946, transferred to RAF in 1952.
Now lives with his wife, a former WAAF, in Cambridge.

SERVICE	PROMOTIONS	AWARDS
Joined RAAF 1940		
Trainee 11 Dec 1940		
To Canada 22 Feb 1941	Sgt 1 Sep 1941	
To England 15 Sep 1941		
106 Sqdn 1941-1942	Plt Off 14 Nov 1942	DFM Gaz
617 Sqdn Mar 1943	F/O 14 Sep 1943	20 Apl 1943 (27 ops)
	F/Lt 14 Nov 1944	
205 Sqdn Korea 1952		
205 Sqdn and Borneo 1955		
209 Sqdn Malaya 1955		

Flight Lieutenant Harold Brownlow Morgan Martin (68795) Pilot:
b 27 Feb 1918, Edgecliffe, Australia.
Now with Hawker Siddeley International, and President of the Bomber Command Association.

Joined RAF 28 Aug 1940		
11 SFT 17 Jun 1941	Plt Off 17 Jun 1941	
No. 2 S of AN 21 Jun 1941		
No. 14 OTU 2 Aug 1941		
455 Sqdn 25 Oct 1941		
50 Sqdn 24 Apl 1942	Fg Off 17 Jun 1942	DFC Gaz
No. 1654 CU 11 Oct 1942		6 Nov 1942 (25 ops)
100 Gp HQ 21 Mar 1943	A/Fl Lt	
617 Sqdn 31 Mar 1943	F/Lt 18 Jun 1943	DSO Gaz
		28 May 1943 (36 ops)
515 Sqdn 1944	Sdn Ldr 1 Sep 1947	Bar DFC 12 Nov 1943
HQ M/East 19 Oct 1944	Wg Cdr 1 Jul 1954	Bar DSO
		31 Mar 1944 (49 ops)
100 Gp HQ 23 Mar 1945	Gr Cpt 1 Jul 1959	2 Bar DFC
No. 1 Ferry Unit Jul 1945	Air Cdre 1 Jan 1963	14 Nov 1944 (83 ops)
295 Sqdn 1 Oct 1945	AVM 1 Jan 1966	
299 Sqdn 15 Jan 1946	AM 1 July 1970	
242 Sqdn 27 Jan 1947		
46 Gp HQ 24 Mar 1947		
Air Ministry 1 Oct 1948		AFC Gaz
Air Attache at Tel Aviv 20 Jun 1952		1 Jan 1949
NATO Paris 26 May 1959		
HG Sig Command 23 Mar 1959		
C in C Cyprus 1961-1962		
HQ Trans Cmd 6 Aug 1962		
HQ 38 Gp 15 Oct 1962		
ADC to Queen 1963		

Flight Lieutenant Harold Brandon Morgan Martin (68795) Pilot: (*continued*)

SERVICE	PROMOTIONS	AWARDS
SASO Near East 1965		
ADC to Queen 1967-1970		CB 1968
RAF HQ 24 Jun 1970		
HQ RAF Germany and C in C 10 Nov 1970		KCB Gaz
Air Member for Personnel 25 Apl 1973		1 Jan 1971
Retired 31 Oct 1974		

Pilot Officer Ivan Whittaker (51740) Flight Engineer:
Died in RAF Halton hospital 1979.

Joined RAF 1938		
No. 1 STT 18 Jan 1938	(No. 573165)	
RAF Cosford 4 Aug 1938		
213 Sqdn 18 Aug 1940		
RAF Ouston 27 Jun 1941		
RAF Catterick 31 Aug 1941		
ACRC 19 Jan 1942		
No. 60 OTU 8 Feb 1942		
No. 10 AGS 21 Mar 1942		
97 Sqdn 24 Apl 1942		
50 Sqdn May 1942		
No. 1654 CU 11 Nov 1942	Plt Off 5 Apl 1943	
617 Squadron 31 Mar 1943	Fg Off 5 Oct 1943	DFC Gaz
		19 Nov 1943
	Wounded 12 Feb 1944	Bar DFC
No. 4 STT 26 May 1944	Fl Lt 5 Apl 1945	Mar 1944 (43 ops)
Eng Branch 9 June 1949	Sdn Ldr 1 Jan 1954	
Tech Branch 1 Dec 1949	Wg Cdr 1 Jul 1959	OBE Gaz
	Gp Cpt 1 Jul 1965	2 June 1962
Retired 19 Oct 1974		

Flight Lieutenant Jack Frederick Leggo RAAF (A/492367) Navigator:
b 21 Apl 1916, Sydney, Australia.   Bank clerk.
Still living in Australia.

Joined RAAF 1940		
Trainee 19 Aug 1940		
To Canada 23 Jan 1941	Sgt 23 Jun 1941	
To England 6 Aug 1941	Later Commission	
455 Sqdn		
50 Sqdn 24 Apl 1942		DFC Gaz
92 Group 16 Sep 1942		6 Nov 1942 (26 ops)
617 Sqdn Mar 1943	Navigation Officer	Bar DFC
Pilot Training 11 Sept 1943		28 May 1943 (35 ops)
10 Sqdn (Aust)	Flew 12 missions (pilot)	
Returned to Austraia 17 Sep 1945		

Flying Officer Leonard Chambers RNZAF (NZ/403758) Wireless Operator:
b 18 Feb 1919, Karramea, New Zealand.   Carpenter.
Now living in Westpoint N.Z.

SERVICE	PROMOTIONS	AWARDS
Joined RNZAF 1940		
75 NZ Sqdn	Flew 31 operations	
No. 26 OTU		
617 Sqdn 7 Apl 1943	Commission	DFC Gaz
Returned to NZ 3 Nov 1944		28 May 1943
Discharged 10 Feb 1945		

Flight Lieutenant Robert Claude Hay RAAF (A/407071) Bomb aimer:
b. 3 Nov 1913, Gawlor, Australia.
Killed in Action 12 February 1944 and is buried at St Michaels Cemetery,
    Cagliari, Sardinia.

Joined RAAF 1940		
Trainee 11 Oct 1940		
To Canada 28 Nov 1940	Sgt 23 Jun 1941	
To England 6 Jul 1941	F/Sgt 23 Dec 1941	
455 Sqdn		
50 Sqdn 24 Apl 1942	Plt Off 16 Dec 1942	DFC Gaz
	F/O 16 June 1942	20 Oct 1942
617 Sqdn Apl 1943	A/Fl Lt – Bombing Leader Bar DFC	
	F/Lt 23 July 1943	

Pilot Officer Bertie Towner Foxlee RAAF (A/404595) Front Gunner:
Queensland, Australia.
Now living in Queensland, Australia.

Joined RAAF 1940		
Trainee 11 Oct 1940		
To Canada 28 Nov 1940	Sgt 23 Jun 1941	
To England 6 Jul 1940	F/Sgt 23 Dec 1941	
455 Sqdn	Plt Off 16 Dec 1942	
50 Sqdn 24 Apl 1942	Fg Off 16 June 1943	DFM Gaz
		4 Jan 1943
617 Sqdn Apl 1943	Fl Lt 16 Dec 1944	DFC Gaz
Returned to Australia, 25 Jul 1944		14 Apl 1944 (8 ops

Flight Sergeant Thomas Drayton Simpson RAAF (A/408076) Rear Gunner:
Hobart, Tasmania.   Law clerk.
Lives in Hobart where he is a partner in a legal firm.

97 Sqdn 10 Oct 1941		
455 Sqdn 20 Feb 1942		
50 Sqdn 7 May 1942	Flew 37 operations	
to ? 28 Oct 1942		
617 Sqdn 6 Apl 1943	Plt Off 19 May 1943	DFM Gaz
		28 May 1943
No. 27 OTU 5 May 1944	Fg Off 28 Dec 1943	DFC Gaz
Pilot's course 17 Oct 1943		14 Apl 1944 (50 op

Squadron Leader Henry Melvin Young (72478) Pilot and A Flight Commander:
b 1916 Pasadena, California, USA.
Killed in Action 17 May 1943.

SERVICE	PROMOTIONS	AWARDS
Joined RAFVR 25 Aug 1939		
No. 1 ITW 25 Aug 1939	Plt Off 25 August 1939	
No. 9 FTC 7 Oct 1939	Fg Off 13 Mar 1940	
RAF Abingdon 6 Apl 1940	Fl Lt 6 August 1941	
102 Sqdn 10 Jun 1940	Sqdn Ldr 1 June 1942	DFC Gaz
No. 10 OTU 7 Feb 1941		9 May 1941 (28 ops)
No. 22 OTU 30 Apl 1941		
104 Sqdn 4 Sep 1941	Middle East	
HQ 205 Gp 29 Jun 1942	(with 104 Sqdn)	Bar DFC
Special Duties 15 Oct 1942	(USA)	18 Sep 1942 (51 ops)
57 Sqdn 13 Mar 1943		
617 Sqn 10 Apl 1943		

Sergeant David Taylor Horsfall (568924) Flight Engineer:
b 1920, Hove, Sussex.
Killed in Action 17 May 1943.

Boy entrant 1936
RAF Halton 14 Jan 1936
97 Sqdn 9 Jan 1939
245 Sqdn 30 Oct 1939
RAF Hendon 6 Apl 1940
No. 1 Sig Dpt 21 Jan 1941
133 Sqdn 7 Nov 1941
No. 4 STT 28 Jul 1942
No. 1654 CU 18 Sep 1942
106 Sqdn 23 Sep 1942
No. 44 CU 8 Nov 1942
No. 1661 CU 9 Nov 1942
RAF Swinderby 15 Dec 1942
57 Sqdn 13 Mar 1943
617 Sqdn 25 Mar 1943

Sergeant Charles Walpole Roberts (126944) Navigator:
b Cromer, Norfolk.   Trainee accountant.
Killed in Action 17 May 1943.

Joined RAF 1940
No. 1 RC 17 Oct 1940
No. 9 RC 26 Oct 1940
A & EE 29 Nov 1940
No. 9 RW 4 Jan 1941
No. 6 ITW 15 Mar 1941
Rhodesia 31 Jul 1941   (Bulawayo)
No. 28 EFTS 28 Aug 1941
No. 75 AS 22 Sep 1941
No. 47 AS 7 Nov 1941

Sergeant Charles Walpole Roberts (126944) Navigator: *(continued)*

SERVICE	PROMOTIONS	AWARDS
No. 43 AS 21 Feb 1942		
No. 3 PRC (UK) 14 Apl 1942		
No. 10 OTU 14 Jul 1942		
No. 9 Flight 8 Nov 1942		
No. 1661 CU 9 Nov 1942		
No. 1660 CU 15 Dec 1942		
57 Sqdn 13 Mar 1943		
617 Sqdn 25 Mar 1943		

Sergeant Lawrence William Nichols (1377941) Wireless Operator:
b 1910, Kent.
Killed in Action 17 May 1943.
  Joined RAF 1940
  No. 2 Sig Sch 13 Sep 1940
  RAF Hendon 8 Nov 1940
  No. 10 SRC
  No. 2 Sig Sch 31 May 1941
  No. 7 ITS 5 Sep 1941
  No. 1 AGS 30 May 1942
  No. 10 OTU 14 Jul 1942
  No. 9 CF 8 Nov 1942
  No. 1661 CU 9 Nov 1942
  57 Sqdn 13 Mar 1943
  617 Sqdn 25 Mar 1943

Flying Officer Vincent Sandford MacCausland RCAF (J/ 15309) Bomb aimer:
b 1913.
Killed in Action 17 May 1943.
  Joined RCAF 1940
  Trainee 13 May 1940
  617 Sqdn 16 Apl 1943

Sergeant Gordon Arthur Yeo (1317656) Front Gunner:
b 1923 South Wales.
Killed in Action 17 May 1943.
  Joined RAF 1941
  Oxford 23 April 1941
  Reserve 24 Apl 1941
  No. 1 ACRC 18 Aug 1941
  No. 12 ITW 6 Sep 1941
  No. 32 EFTS 29 Oct 1941
  No. 36 FTS
  No. 31 PD   (New Brunswick Canada)
  No. 3 PRC 24 Mar 1942
  No. 11 STT 15 Jul 1942
  No. 1 AAS 26 Sep 1942
  No. 1661 CU 9 Nov 1942
  RAF Swinderby 15 Dec 1942
  57 Sqdn 13 Mar 1943
  617 Sqdn 25 Mar 1943

Sergeant Wilfred Ibbotson (655431) Rear Gunner:
b 1914 Brelton, West Yorks.
Killed in Action 17 May 1943.

SERVICE	PROMOTIONS	AWARDS
Joined RAF 1941	·	
No. 9 RW 29 Mar 1941		
No. 11 ITW 12 Jul 1941		
No. 10 (S) RC 12 Jun 1941		
RDU 19 Aug 1941		
Uxbridge 12 Oct 1941		
Acklington 18 Nov 1941		
ACRC 2 Jan 1942		
No. 14 ITW 10 Jan 1942		
No. 4 AGS 6 Jun 1942		
No. 10 OTU 28 Jul 1942		
No. 9 CF 8 Nov 1942		
No. 1661 CU 9 Nov 1942		
No. 1660 CU 15 Feb 1943		
57 Sqdn 13 Mar 1943		
617 Sqdn 23 Mar 1943		

Flight Lieutenant William Astell (60283) Pilot:
b 1920 Manchester.
Killed in Action 17 May 1943.

Joined RAFVR 1939		
No. 20 FTS 3 Nov 1940	(Rhodesia)	
No. 70 OTU 5 Nov 1940	Plt Off 3rd Nov 1940	
148 Sqdn 30 Jan 1941	Middle East	DFC Gaz
No. 1654 CU 4 Oct 1942	Fg Off 3 Nov 1941	14 Aug 1942
No. 1485 TT & GF 3 Dec 1942	(England)	
57 Sqdn 26 Jan 1943	F/Lt 3 Nov 1942	
617 Sqdn 25 Mar 1943		

Sergeant John Kinnear (635123) Flight Engineer:
b 1922 Fife, Scotland.   Garage hand.
Killed in Action 17 May 1943.

Joined RAF 1939
RAF Driffield 20 Feb 1939
RAF Cardington 29 Jun 1939
RAF St Athan 7 Jul 1939
No. 8 FTS 12 Jan 1940
No. 2 STT 14 Nov 1940
No. 55 OC 15 Jan 1941
No. 55 OTU
No. 4 STT 1 Aug 1942
No. 1654 CU 12 Oct 1942
617 Sqdn 25 Mar 1943

Pilot Officer Floyd Alwin Wile RCAF (J/18872) Navigator:
b 1919 Truro, Nova Scotia, Canada.
Killed in Action 17 May 1943.

Joined RCAF 9 May 1941	Commission
617 Sqdn	25 Mar 1943

Sergeant Abram Garshowitz RCAF (R/84377) Wireless Operator:
Ontario, Canada.
Killed in Action 17 May 1943.

SERVICE	PROMOTIONS	AWARDS
Joined RCAF 28 Jan 1941		
617 Sqdn 26 Mar 1943		

Flying Officer Donald Hopkinson (127817) Bomb aimer:
b 19 Sep 1920 Royton, Oldham, Lancs.   Clerk.
Killed in Action 17 May 1943.

Joined RAF 1941		
No. 3 RC 13 Feb 1941	(LAC No. 1036687)	
No. 1 RW 28 Jun 1941		
ITW 5 Jul 1941		
No. 1 Depot		
No. 13 EFTS 24 Sep 1941	(Canada)	
CTS 9 Nov 1941		
No. 7 AOS 5 Dec 1941		
No. 3 B & GS 14 Mar 1942	Plt Off 25 May 1942	
57 Sqdn	Fg Off 25 May 1942	
617 Sqdn		

Sergeant Francis Anthony Garbas RCAF (R/103201) Front Gunner:
Ontario, Canada.
Killed in Action 17 May 1943.

Joined RCAF 22 May 1941
617 Sqdn 25 Mar 1943

Sergeant Richard Bolitho (1211045) Rear Gunner:
b 1920 Kimberly, Notts, but originally from Co Antrim, N. Ireland.
Killed in Action 17 May 1943.

Joined RAF 1940
No. 2 RC 25 Nov 1940
No. 4 RC 9 Dec 1940
RAF Calshot 10 Jan 1941
No. 1 RW 26 Apl 1941
No. 7 ITW 10 May 1941
51 Gp Pool 28 Jun 1941
ADRC 5 Aug 1941
RDU 6 Oct 1941
No. 14 ITW 10 Jan 1942
No. 9 AGS 5 Jun 1942
No. 19 OTU 27 Aug 1942
No. 1654 CU 15 Oct 1942
9 Sqdn 23 Dec 1942
27 Sqdn 22 Jan 1943
617 Sqdn 25 Mar 1943

Flight Lieutenant David John Maltby (60335) Pilot:
b 1920 Hastings, Sussex.   Engineering Student.
Killed in Operations 14 September 1943.

SERVICE	PROMOTIONS	AWARDS
Joined RAFVR 1940		
RAF Uxbridge 20 Mar 1940	Plt Off 18 Jan 1941	
No. 4 ITW 24 Jun 1940	Fg Off 12 Jan 1942	
51 Gp Pool 19 Aug 1940	Fl Lt 12 Jan 1943	
No. 12 SFTS 28 Sep 1940		
No. 2 Nav Sch 18 Jan 1941		
No. 16 OTU 15 Mar 1941		
106 Sqdn 4 Jun 1941		
No. 1654 CU 14 Jun 1942		
97 Sqdn 26 Jun 1942		DFC Gaz
No. 1485 TTGU 14 Jul 1942		11 Aug 1942 (27 ops)
97 Sqdn 15 Mar 1943		
617 Sqdn 25 Mar 1943		DSO Gaz
		28 May 1943

Sergeant William Hatton (1013557) Flight Engineer:
b 1921 Wakefield, Yorks.
Killed on Operations 14 September 1943.

Joined RAF 1940
RAF Padgate 1 Aug 1940
Bo. 3 RC 3 Aug 1940
No. 10 STT
111 Sqdn 31 Dec 1940
No. 2 STT 21 Feb 1941
RAF Speke 1 May 1941
MSFU 12 May 1941
No. 45 STT 21 Oct 1942
No. 1660 CU 5 Jan 1943
97 Sqdn 15 Mar 1943
617 Sqdn 25 Mar 1943

Sergeant Vivian Nicholson (1144183) Navigator:
b 1923 Sherborn, Co. Durham.   Joiner's Apprentice.
Killed on Operations 14 September 1943.

Joined RAF 1941
No. 3 RC 18 Feb 1941
Reserve 19 Feb 1941
No. 1 ARC 7 Jul 1941
No. 1 ITW 26 Jul 1941
To Canada
No. 9 ADS 17 Jan 1942
No. 10 AFU
No. 10 OTU 22 Sep 1942
No. 1660 CU 5 Jan 1943
97 Sqdn 18 Mar 1943
617 Sqdn 25 Mar 1943

Sergeant Anthony Joseph Stone (1311959) Wireless Operator:
b 1921 Winchester.
Killed on Operations 14 September 1943.

SERVICE	PROMOTIONS	AWARDS
Joined RAF 1940		
Uxbridge 4 Nov 1940		
No. 4 RC 26 Nov 1940		
Thorney Is 27 Dec 1940		
No. 10 SRC 6 Jun 1941		
No. 2 Sig Sch 8 Aug 1941		
No. 11 OTU 7 Nov 1941		
No. 1 AGS 8 Aug 1942		
No. 100 TU		
No. 1660 CU 5 Jan 1943		
207 Sqdn 17 Feb 1943		
97 Sqdn 15 Mar 1943		
617 Sqdn 25 Mar 1943		

Pilot Officer John Fort (477575) Bomb aimer:
b York.
Killed on Operations 14 September 1943.

Joined RAF 1929		
RAF Halton 15 Jan 1929		
CFS 5 Jan 1932		
HMS *Glorious* 23 Jul 1935		
504 Sqdn 5 Nov 1937		
RAF Catterick 22 Apl 1940		
41 Sqdn 8 Apl 1940		
ACRC 1 Dec 1941		
ACDW 3 Jan 1942		
No. 9 ITW 24 Jan 1942		
No. 51 Gp Pool 20 May 1942		
No. 10 AFU 24 Jun 1942		
No. 10 OTU 24 Sep 1942	Plt Off 14 Sep 1942	
No. 1660 CU 5 Jan 1943		
207 Sqdn 17 Feb 1943		
97 Sqdn 15 Mar 1943		
617 Sqdn 25 Mar 1943	Fg Off 20 Aug 1943	DFC Gaz 28 May 1943 (2 op

Flight Sergeant Victor Hill (1315725) Front Gunner:
Killed on Operations 14 September 1943.

Joined RAF 1941		
RAF Blackpool 10 Mar 1941		
No. 8 RC 27 Mar 1941		
RAF St Eval 20 Apl 1941		
No. 10 (S) RC 4 Sep 1941		
RDU 8 Dec 1941		

Flight Sergeant Victor Hill (1315725) Front Gunner: (continued)

SERVICE	PROMOTIONS	AWARDS
West Malling 24 Dec 1941		
ACRC 2 Jun 1942		
No. 14 ITW 3 Jun 1942		
No. 1 AAS 5 Jul 1942		
9 Sqdn 15 Aug 1942		
617 Sqdn 7 May 1943		

Sergeant Harold Thomas Simmonds (1248156) Rear Gunner:
b 1921 Burgess Hill, Sussex.
Killed on Operations 14 September 1943.

Joined RAF 1941
No. 2 RC 22 Mar 1941
No. 3 RC 22 Sep 1941
No. 10 (S) RC 25 Sep 1941
R & DU 31 Dec 1941
RAF Kemble 5 Jan 1942
Mount Batten 29 May 1942
No. 11 STT 9 Sep 1942
No. 2 AGS 6 Nov 1942
No. 1660 CU 5 Jan 1943
97 Sqdn 15 Mar 1943
617 Sqdn 25 Mar 1943

Squadron Leader Henry Eric Maudsley (62275) Pilot and B Flight Commander:
b 1912, lived in Sherbourne, Notts.   Eton College.
Killed in Action 17 May 1943.

Joined RAFVR 1940		
No. 2 RC 3 Apl 1940		
No. 1 RC 20 Jun 1940		
No. 5 ITW		
No. 51 Gp Pool 19 Aug 1940		
No. 10 SFTS 28 Sep 1940		
No 10 SFTS 1 Nov 1940	Plt Off 29 Jan 1941	
44 Sqdn 22 May 1941		DFC Gaz
No. 1654 CU 2 Oct 1941	Fg Off 29 Jan 1942	30 Jan 1942 (28 ops)
50 Sqdn 1 Mar 1943	Fl Lt 29 Jan 1943	
617 Sqdn 25 Mar 1943		

Sergeant John Marriott (1003474) Flight Engineer:
b 19 Jan 1920 New Smithy, Buxton, Derbyshire.   Factory worker.
Killed in Action 17 May 1943.

Joined RAF 1940
RAF Padgate 7 Jun 1940
No. 39 MU 25 Jun 1940
RAF Locking 19 Oct 1940

Sergeant John Marriott (1003474) Flight Engineer: (continued)

SERVICE	PROMOTIONS	AWARDS
18 Sqdn 13 Mar 1941		
No. 2 STT 11 Apl 1941		
No 4 STT 7 Jul 1942		
50 Sqdn 25 Aug 1942		DFM Gaz
617 Sqdn 25 Mar 1943		19 Jun 1943 (27 ops)

Flying Officer Robert Alexander Urquhart RCAF (J/9763) Navigator:
Killed in Action 17 May 1943.

Joined RCAF 1941		
Trainee 9 Jan 1941	Commission	
50 Sqdn 23 Sep 1942	Wounded 17 Dec 1942	DFC Gaz
		20 Jul 1945 (28 ops)
617 Sqdn 25 Mar 1943		wef 18 May 1943

Sergeant Allan Preston Cottam RCAF (J/93558) Wireless Operator:
b 1913 Alberta, Canada.
Killed in Action 17 May 1943.
    Joined RCAF 1941
    Trainee 22 Mar 1941
    617 Sqdn 25 Mar 1943

Pilot Officer Michael John David Fuller (143760) Bomb aimer:
b 1920 West Wickham, Kent.
Killed in Action 17 May 1943.

Joined RAFVR 1940		
No. 2 RC 8 May 1940		
No. 1 ITW 13 June 1940		
No. 47 AS 9 Feb 1941		
No. 75 AS 16 Jun 1941		
No. 65 AS 23 Jun 1941		
No. 3 PRC 18 Oct 1941		
OAFU 24 Apl 1942		
No. 13 OTU 28 Apl 1942		
No. 25 OTU 13 Jul 1942		
No. 1654 CU 31 Oct 1942		
106 Sqdn 24 Dec 1942		
50 Sqdn 12 Feb 1943	Plt Off 9 Mar 1943	
617 Sqdn 25 Mar 1943		

Flying Officer William John Tytherleigh (120851) Front Gunner:
b 1922, Hove, Sussex.
Killed in Action 17 May 1943.
    Joined RAF 1940
    No. 1 RC 25 Apl 1940
    No. 4 RC 19 Jul 1940
    No. 10 RC 16 Aug 1940
    No. 4 W & SS 15 Nov 1940

Flying Officer William John Tytherleigh (129851) Front Gunner:

SERVICE	PROMOTIONS	AWARDS
No. 58 OTU 3 Feb 1941		
No. 7 B & GS 12 Apl 1941		
50 Sqdn 20 Apl 1941	Sgt 923074	
No. 16 OTU 25 Apl 1941		
50 Sqdn 16 Aug 1941		DFC Gaz
		29 Jun 1945 (42 ops)
RAF Swinderby 18 Apl 1942	Plt Off 20 Apl 1942	wef 16 May 1943
No. 1654 CU 23 Jul 1942	Fg Off 20 Oct 1942	
617 Sqdn 25 Mar 1943		

Sergeant Norman Rupert Burrows (1503194.4) Rear Gunner:
Killed in Action 17 May 1943.
  Joined RAF 1941
  No. 3 RC 14 Jun 1941
  Reserve 15 Jun 1941
  No. 3 RC 16 Oct 1941
  No. 10 (S) RC 20 Oct 1941
  RDU 12 Jan 1942
  RAF Speke 3 Feb 1942
  No. 14 ITW 22 Jun 1942
  No. 2 AGS 17 Jul 1942
  No. 1654 CU 29 Aug 1942
  50 Sqdn 30 Sep 1942
  617 Sqdn 25 Mar 1943

Flight Lieutenant Leslie Gordon Knight RAAF (A/401449) Pilot:
b 7 Mar 1921 Camberwell, Victoria Australia.   Student Accountant.
Killed in Action 16 September 1943.

Joined RAAF 1941		
Trainee 3 Feb 1941		
To England 17 Nov 1941	Sgt 17 Oct 1941	
50 Sqdn 22 Sep 1942	Plt Off 8 Dec 1942	
No. 1654 CU Mar 1943	Flew 26 operations	
	F/O 8 June 1943	
617 Sqdn Mar 1943	A/Flt Lt 1 Aug 1943	DSO Gaz
		28 May 1943
		MinD (Post)
		1 Jan 1945

Sergeant Raymond Ernest Grayston (913888) Flight Engineer:
Prisoner of War 16 September 1943 and sent to Stalag Luft III.
Returned to UK 27 May 1945; Released from RAF 30 Oct 1945.
Living in Sussex.
  Joined RAFVR 1940
  RAF Uxbridge 8 Feb 1940
  No. 7 RC 13 Feb 1940
  No. 3 STT 8 Mar 1940

Sergeant Raymond Ernest Grayston (913888) Flight Engineer: (*continued*)

SERVICE	PROMOTIONS	AWARDS
No 6 FTS 7 Aug 1940		
No. 3 STT 8 Aug 1941		
No. 32 MU 11 Mar 1942		
No. 4 STT 27 Aug 1942		
50 Sqdn 10 Oct 1942		
617 Sqdn 25 Mar 1943		

Flying Officer Harold Sidney Hobday (119291) Navigator:
b Croydon, Surrey.
Now lives in Sussex.

Joined RAFVR 1940		
No. 9 RC 3 Sep 1940		
N.45 AONS 11 Sep 1940		
St Eval 8 Nov 1940		
No. 1 RW 1 Feb 1943		
No. 4 ITW 1 Mar 1941		
No. 42 Air Sch 10 Jan 1942	South Africa – Plt Off	
No. 2 AFU 7 Apl 1942	10 Jan 1942	
No. 14 OTU 5 May 1942		
No. 1654 CU 13 Aug 1942		
50 Sqdn 19 Sep 1942	Flew 26 operations.	
	F/O 10 Oct 1942	
617 Sqdn 25 Mar 1943	Fl Lt 10 Jan 1944	DFC Gaz
24 Sqdn 28 Feb 1944		28 May 1943
Released 25 Apl 1946		

Sergeant Robert George Thomas Kellow RAAF (A/411453) Wireless Operator:
b 13 Dec 1916 Newcastle, NSW, Australia.   Shop Assistant.
Shot down 16 September 1943, evaded via Holland, France, Spain and Gibraltar,
arrived back in UK 4 Dec, 1943;
Returned to Australia 9 May 1944.

Joined RAAF 1941		
Trainee 28 Apl 1941		
To England 13 Jun 1941		
50 Sqdn		DFM Gaz
No. 1654 CU		15 Jun 1943 (24 ops)
617 Sqdn Mar 1943	Plt Off 11 Mar 1943	
	Fg Off 12 Sep 1943	
	Fl Lt 23 Mar 1945	

Flying Officer Edward Cuthbert Johnson (119126) Bomb aimer:
b 1912.
Shot down 16 September 1943, evaded and returned to England via Holland, France,
Spain, Gibraltar in November 1943.
Arrived UK December 1943, now living in Blackpool.

Joined RAFVR 1940		
No. 3 RC 24 Jun 1940	No 1006393	
RAF Debden 19 Jul 1940		
No. 10 FTS 15 Aug 1940		

Flying Officer Edward Cuthbert Johnson (119126) Bomb aimer: (continued)

SERVICE	PROMOTIONS	AWARDS
No. 5 ETS Nov 1940		
RAF Padgate 27 Dec 1940		
No. 1 RW 18 Feb 1941		
N Depot 17 Jul 1941		
No. 4 A & G 4 Aug 1941		
No. 4 BGS 26 Oct 1941		
No. 2 ANS 6 Dec 1941	Plt Off 11 Jan 1942	
No. 2 AFU 7 Apl 1942		
106 Sqdn 21 Aug 1942		
No. 1654 CU 16 Sep 1942		
50 Sqdn 18 Sep 1942	Fg Off 4 Oct 1942	
617 Sqdn 25 Mar 1943		DFC Gaz
Uxbridge 15 Dec 1943	Fl Lt 10 Jan 1944	28 May 1943 (23 ops)
West Freugh 27 Jan 1944		
Released 17 Jul 1947		

Sergeant Frederick E. Sutherland RCAF (R/108628) Front Gunner:
Shot down 16 September 1943, evaded and returned to England via Holland, France, Spain.
He lives in Canada.
No. 1660 CU Apl 1942
50 Sqdn Sep 1942
617 Sqdn 29 Mar 1943

Sergeant Harry E. O'Brien RCAF (R142252) Rear Gunner:
Prisoner of war 16 September 1942, sent to POW Camp No. 357 Kopernikus.
Now living in Canada.
No. 1654 CU 1942
617 Sqdn 25 Mar 1943

Flight Lieutenant David John Shannon RAAF (A/407729) Pilot:
b 27 May 1922 St Umley Park, South Australia.   Clerk.
Now a business man in London.

Joined RAAF 1941		
Trainee 4 Jan 1941		
Canada 1941	Plt Off 22 Sep 1941	
	F/O 23 Mar 1942	
106 Sqdn 1942	Flew 36 operations	DFC Gaz
		12 Jan 1943 (26 ops)
617 Sqdn 1943	Fl Lt 23 Sept 1043	DSO Gaz
		28 May 1943
		Bar DFC 12 Nov 1943
OTU 25 July 1944		Bar DSO
		26 Sep 1944 (65 ops)
511 Sqdn		
246 Sqdn	S/L 1st Jan 1945	
Returned Home 15 Dec 1945		

Sergeant Robert Jack Henderson (544401) Flight Engineer:
From Ayr, Scotland.

SERVICE	PROMOTIONS	AWARDS
Joined RAFVR 1937		
Uxbridge 21 Sep 1937		
Cardington 2 Oct 1937		
No. 3 STT 7 Jan 1938		
Cranwell 1 Mar 1938		
78 Sqdn 20 Mar 1940		
207 Sqdn 31 Dec 1940		
No. 2 STT 21 Nov 1941		
No. 11 RC 25 Mar 1942		
207 Sqdn 22 Apl 1942	Warrant Officer	
No. 4 STT 18 Jul 1942		
57 Sqdn 10 Oct 1942	Flew 16 operations	
617 Sqdn 4 Apl 1943	Plt Off 5 Apl 1944	DFM Gaz
No. 1654 CU 5 Aug 1944	Fg Off 5 Oct 1944	30 Jun 1944 (33 ops)
Discharged 5 May 1948	Fl Lt 5 Apl 1946	
Re-enlisted 27 Jul 1948		
203 Sqdn 6 Aug 1948		
St Eval 2 Dec 1948		
No. 5 PDL 28 Jul 1949		
Rhodesia 4 Aug 1949		
No. 3 ANS 4 Aug 1949		
No. 4 FTS 9 Oct 1950		
Coltishall 25 Apl 1952		
No. 238 OCU 22 Mar 1954		
Turnhouse 8 Feb 1956		
Rufforth 1 Jan 1957		
AMU Ruislip 1 Dec 1958		
Akorotiri 1 Mar 1959	Cyprus	
249 Squadron 13 Jun 1960		
RAF Luqa 18 Sep 1960	Malta	

Flying Officer Daniel Revie Walker RCAF (J/15336) Navigator:
b 1917 Blairmore, Alberta.   Forest Officer's Assistant.
Now living in Canada.

Joined RCAF 1940		
RCAF Reserve Aug 1940		
106 Sqdn Apl 1942	Flew 32 operations	DFC Gaz
No. 22 OTU Dec 1942		12 Jan 1943 (30 ops)
54 Base HQ Apl 1944		Bar 28 May 1943
No. 2 AC HQ Jan 1945		
No. 7 Release Ctr Apl 1946		
No. 5 Release Ctr Aug 1946		
No. 2 AC HQ Sep 1946		
12 Gp HQ Oct 1947		
Vancouver Jun 1948		
12 Gp HQ Aug 1948		

Flying Officer Daniel Revie Walker RCAF (J/15336) Navigator: (continued)

SERVICE	PROMOTIONS	AWARDS
Maxwell AFB Jul 1949	USAF	
Ottawa Dec 1949	Wg Cdr Jan 1952	
Air Force HQ Aug 1952		
No. 2 ANS Aug 1954	Officer Commanding	
Air Force HQ Jul 1957		
McCord AFB Jan 1960	Washington	
Summerside Stn Apl 1966	Chief Admin Off	
Can Forces Base Apl 1966		
Can Forces HQ Nov 1966		
Released 1967		

Flying Officer Concave Brian Goodale (101042) Wireless Operator:
b 12 June 1919.
Died in Cambridge Hospital, December 1977.

Joined RAFVR 1939		
No. 2 Depot 21 Oct 1939		
No. 2 E & WS 8 Dec 1939		
No. 7 B & GS 15 Jun 1940		
No. 10 OTU 6 Jul 1940		
51 Sqdn 9 Sep 1940	Plt Off 18 Jul 1941	DFC Gaz
RAF Disforth 17 Jan 1942		30 Jan 1942 (28 ops)
No. 1652 CU 1 Feb 1942	Fg Off 18 Jul 1942	
617 Sqdn 20 Apl 1943		Bar DFC Gaz
No. 100 OTU 22 Apl 1944		28 May 1943
HQ Tech Tr Cd 11 May 1944		
No. 105 OTU 4 Sep 1944	Fl Lt 1 Sep 1945	
HQ 45 Gp 2 Nov 1945		
Sec BOAC 29 Nov 1945		
HQ Tr Cmnd 15 Dec 1945		
224 Sqdn 17 Dec 1945		
HQ FTC 10 Nov 1947	Sqn Ldr 1 Aug 1947	
Debden 20 Oct 1948		
Watton 8 Nov 1948		
Henlow 21 Sep 1950		
Debden 12 Nov 1951		
Trimley Heath 14 Feb 1952		
783 Sig Unit 26 Feb 1952		
HQ B/Cmnd 30 Dec 1953		
330 Sig Unit 5 Jan 1954		
266 Sig Unit 5 Jan 1954		
Swanton Morley 21 Aug 1958		
151 Wing 26 Sep 1960		
Retired 14 Jun 1961	Sqdn Ldr	

Flight Sergeant Leonard Joseph Sumpter (655673) Bomb aimer:
b 1911 Kettering, Northants.
Grenadier Guards in 1928-31, 1939-41.
Released 11 March 1946, rejoined 1946-1950, Physical Fitness Branch.
Living in Luton.

SERVICE	PROMOTIONS	AWARDS
Trans RAF 1941		
No. 9 RW 12 Apl 1941		
No. 10 ITW 26 Jun 1941		
No. 33 ANS 15 Sep 1941		
Charlottown 16 Sep 1941		
Nom 31 B & G Sch 14 Feb 1942		
No. 3 PRC 13 May 1942		
No. 11 OTU 27 Jun 1942		
57 Sqdn 8 Sep 1942	Flew 13 operations	
617 Sqdn Mar 1943	Plt Off 27 Jun 1943	DFM Gaz 28 May 1943
85 OTU 21 Aug 1944	Fg Off 27 Dec 1943	DFC Gaz 9 Jun 1944
617 Sqdn 22 Jan 1945	Fl Lt 17 Jan 1945	(35 ops)
RAF Burn 11 Sep 1945		
18 Gp 11 Mar 1946		

Sergeant Brian Jagger (1320873) Front Gunner:
b 1922 Chelsea, London.
Killed in a flying accident 30 April 1944, in a Lancaster of 49 Squadron. On a fighter Affiliation Exercise, a dinghy became released and wrapped itself around the tailplane causing the aircraft to crash. He is buried in Cambridge City Cemetery.

Joined RAF 1941		
Enlisted 26 Jun 1941		
No. 3 RC 20 Oct 194	Fl Sgt	
No. 10 (S) RC 22 Oct 1941	Plt Off 23 Oct 1943	
No. 26 OTU 15 Jan 1942		
ASRC 2 Jun 1942		
No. 14 ITW 6 Jun 1942		
No. 2 AGS 17 Jul 1942		
No. 1654 CU 29 Aug 1942		
50 Sqdn 29 Sep 1942		
617 Sqdn 29 Apl 1943		DFM Gaz
BDU 6 Mar 1944	Fg Off 23 Mar 1944	19 Sep 1943 (24 ops)

Pilot Officer Jack Buckley (129460) Rear Gunner:
b Bradford, Yorkshire.
Living in Yorkshire.

Joined RAFVR 1939
Depot 6 Sep 1939
Duxford 19 Feb 1940
No. 4 ITW 8 Apl 1940
No. 10 B & GS 4 May 1940
No. 1 SAG 3 Jun 1940
225 Sqdn 23 Jun 1940

Pilot Officer Jack Buckley (129460) Rear Gunner: (continued)

SERVICE	PROMOTIONS	AWARDS
No. 20 OTU 2 Jun 1941		
75 Sqdn 1 Sep 1941	Plt Off 6 Jun 1942	
No. 10 OTU 29 Jun 1942		
75 Sqdn Jun 1942	Flew 35 operations	
No. 10 OTU		
617 Sqdn 6 Apl 1943	Fg Off 6 Jul 1943	DFC Gaz
CGS 18 Aug 1943		28 May 1943
No. 1661 CU 10 Jul 1944	Fl Lt 6 Jun 1944	
No. 57 Base 4 Aug 1944		
RAF Blyton 9 Oct 1945		
Released 25 Jan 1946		

Flight Lieutenant Joseph Charles McCarthy RCAF (J/9346) Pilot:
b 31 Aug 1919 St James Is, New York, USA.

Joined RCAF 1941		
PRC 22 Jan 1942	To England	
No. 12 SFTS		
No. 3 ATC 23 Feb 1942		
No. 14 OTU 19 May 1942	1,000 bomber raids.	
No. 97 CU		
106 Sqdn		
97 Sqdn 22 Sep 1942		DFC Gaz
		14 May 1943 (29 ops)
617 Sqdn Mar 1943		DSO Gaz
		28 May 1943
		Bar DFC
		28 Apl 1944 (48 ops)

Sergeant William Radcliffe (639976) Flight Engineer:
New Gordon, Westminster, British Columbia, Canada.
Killed in a car accident in May 1952, being drowned in the Fraser River, British Columbia.

Joined RAF 1939		
Finningley 13 Apl 1939	As Groundcrew	
No. 1 Wing 24 Jun 1939		
56 Sqdn 15 Dec 1939		
Digby May 1940		
North Weald 11 Feb 1941		
97 Sqdn 4 Sep 1942		
617 Sqdn 25 Mar 1943	Plt Off 24 Sep 1943	DFC Gaz
No. 1654 CU 5 Aug 1944	Fg Off 24 Mar 1944	23 Mary 1944 (36 ops)
Remustered RCAF		

Flight Sergeant Donald Arthur McLean RCAF (J/112723) Navigator:
Toronto, Canda.   School teacher.
Living in Canada.

SERVICE	PROMOTIONS	AWARDS
Joined RCAF 1941		
Enlisted Jun 1941		
Ferry Command Apl 1942		
44 Sqdn		
97 Sqdn	Flew 19 operations	
617 Sqdn Mar 1943	Plt Off 18 Mar 1943	DFM Gaz
No. 1664 CU 20 Jul 1944	Fg Off 20 Oct 1943	28 May 1943
	Flew 52 operations	
Rejoined RCAF post war		
Retired as Wing Commander		

Sergeant Leonard Eaton (110754) Wireless Operator:
Died 1974.

Padgate 22 Jul 1940		
No. 6 RC 25 Jul 1940		
Feltwell 16 Aug 1940		
No. 10 (S) RC 19 Jul 1940		
No. 1 Sig Sch 14 Feb 1941		
No. 7 AACU 7 Jun 1941		
No. 2 Sig Sch 3 Oct 1941		
No. 1 AGS 14 Feb 1942		
No. 14 OTU 7 Apl 1942		
207 Sqdn 9 Aug 1942		
No. 100 CU 15 Sep 1942		
Coningsby 1 Dec 1942		
617 Sqdn 25 Mar 1943	F/Sgt 1 May 1943	DFM Gaz
No. 84 OTW 1 Aug 1944		30 Jun 1944 (35 ops)

Sergeant George Leonard Johnson (1199696) Bomb aimer:
b 1921 nr Horncastle, Lincs.
Living Newark.

Joined RAF 1940
No. 2 RC 6 Nov 1940
No. 7 RC 18 Nov 1940
No. 12 FTS 18 Dec 1940
No. 1 RW 5 Apl 1941
No. 7 ITW 26 Apl 1941
Arnold 6 Jul 1941
Toronto, Can 17 Jul 1941
Arcadia 24 Jul 1941
No. 51 B & GS 10 Sep 1941
No. 51 PD 30 Oct 1941
No. 5 PRC 20 Jan 1942
ACRC 30 Mar 1942
No. 14 ITW 13 Jun 1942

Sergeant George Leonard Johnson (1199696) Bomb aimer: (continued)

SERVICE	PROMOTIONS	AWARDS
No. 1654 CU 26 Jul 1942		
97 Sqdn 27 Jul 1942	Flew 28 operations	
617 Sqdn 25 Mar 1943		DFM Gaz
No. 1 AAS 21 Aug 1943	Plt Off 29 Nov 1943	28 May 1943
No. 1654 CU 20 Mar 1945	Fg Off 29 May 1944	
No. 1656 CU 17 Jul 1945		
Binbrook 1 Oct 1945	Fl Lt 29 Nov 1945	
Manby 1 Apl 1946		
No. 1 ANS 13 Jul 1948		
100 Sqdn 21 Jan 1949		
West Freugh 2 May 1949		
AFS 17 Jan 1950		
236 CU 12 Apl 1950		
120 Sqdn 12 Apl 1950		
St Mawgan 10 Jul 1952		
19 GP HQ 11 Jan 1954	S/Ldr 1 Apl 1955	
Singapore 11 Sep 1957		
Hemswell 24 Aug 1960		
Retired 15 Sep 1962		

Sergeant Ronald Batson (1045069) Front Gunner:
From the North of England

No. 3 RC 29 Mar 1941		
No. 9 RC Nov 1941		
North Weald 6 Jan 1942		
No. 16 RC 21 Jan 1942		
ACRC 22 Jan 1942		
No. 14 ITW		
No. 4 AGS		
No. 47 CF 7 Sep 1942		
No. 106 CF 11 Sep 1942		
97 Sqdn 22 Sep 1942		
617 Sqdn Mar 1943	T/F/Sgt 1 July 1943	DFM Gaz
No. 5 PDC 5 Apl 1945		23 May 1944 (37 ops)
No. 100 PDC 20 Jan 1946		

Flying Officer David Rodger RCAF (10160) Rear Gunner:
b 23 Feb 1918 Sault St Marie, Ontario, Canada.
Now Living in Canada.

Joined RCAF 1941		
Recruit Oct 1941		
To England Mar 1942		
97 Sqdn Oct 1942	Fg Off 19 Nov 1942	
617 Sqdn Mar 1943	Fl Lt 24 Aug 1943	DFC Gaz
To Canada 23 Aug 1944		15 Sep 1944

Sergeant Vernon William Byers RCAF (J/17474) Pilot:
b 1911 Star City, Saskatchewan, Canada.
Killed in Action 16 May 1943.

SERVICE	PROMOTIONS	AWARDS
Joined RCAF 1941		
Recruit 9 May 1941	Commission	
Trained in Manitoba		
467 Sqdn 5 Feb 1943		
617 Sqdn 28 Mar 1943		

Sergeant Alastair James Taylor (575430) Flight Engineer:
b 1923 Alves, Morayshire.   Apprentice.
Killed in Action 16 May 1943.

Joined RAF 1939
RAF Halton 17 Jan 1939
No. 9 FTS 5 Oct 1940
No. 4 STT 14 Aug 1942
No. 1654 CU 3 Nov 1942
467 Sqdn 5 Feb 1943
617 Sqdn 25 Mar 1943

Pilot Officer James Herbert Warner (128619) Navigator:
Killed in Action 16 May 1943.

Joined RAF 1940	
No. 2 RC 16 Dec 1940	
Bridgenorth 23 Dec 1940	
Church Fenton 28 Jan 1941	
No. 1 RW 19 May 1941	LAC No. 1222006
No. 5 ITW 31 May 1941	
ATTS 7 Aug 1941	
Trenton 14 Oct 1941	
No. 31 PD 25 Oct 1941	
No. 7 Depot 13 Dec 1941	
No. 3 PCR 26 Dec 1941	
ACRC 7 Mar 1942	
No. 1 EAOS 28 Mar 1942	
No. 5 AOS 2 Jun 1942	Plt Off 1 Sep 1942
No. 1654 CU 8 Dec 1942	
617 Sqdn 29 Mar 1943	

Sergeant John Wilkinson (102580) Wireless Operator:
b 1922 Antrobus, Cheshire
Killed in Action 16 May 1943.

Joined RAF 1940
Padgate 8 Oct 1940
No. 14 SFTS 16 Nov 1940
No. 10 SRC 10 Apl 1941
No. 2 Sig Sch 8 Aug 1941
RAF Leeming 7 Nov 1941

Sergeant John Wilkinson (102580) Wireless Operator: (*continued*)

SERVICE	PROMOTIONS	AWARDS
No. 2 Sig Sch 17 Jun 1942		
No. 1 AGS 25 Jun 1942		
No. 290 OTU 1 Sep 1942		
No. 1654 CU 8 Dec 1942		
467 Sqdn 5 Feb 1943		
617 Sqdn 25 Mar 1943		

Sergeant Arthur Neville Whitaker (144777) Bomb aimer:
Killed in Action 16 May 1943.
      Enlisted 3 Sep 1939
      No. 9 RW 10 May 1941
      No. 12 ITW 21 May 1941
      No. 4 AOS 6 Sep 1941
      No. 13 OTU 14 Apl 1942
      No. 25 OTU 13 Jul 1942
      No. 49 CF 31 Oct 1942
      467 Sqdn 9 Nov 1942
      617 Sqdn 28 Mar 1943   Plt Off 18 May 1943

Sergeant Charles McAllister Jarvie (1058757) Front Gunner:
b 1922 Glasgow, Scotland.
Killed in Action 16 May 1943.
      Joined RAF 1940
      Padgate 11 Jul 1940
      No. 6 RC 13 Aug 1940
      No. 1 E & WS 30 Aug 1940
      Ullsworth 14 Oct 1940
      Ouston 15 May 1941
      Acklington 29 May 1941
      No. 14 ITW 30 Jul 1942
      No. 4 AGS 6 Sep 1942
      No. 49 CU 31 Oct 1942
      No. 1661 CU
      467 Sqdn 9 Nov 1942
      617 Sqdn 28 Mar 1943

Sergeant James McDowell RCAF (R/ 101749) Rear Gunner:
Killed in Action 16 May 1943.
      Joined RAF 1941
      Enlisted 8 May 1941
      617 Sqdn 28 Mar 1943

Flight Lieutenant Robert Norman George Barlow RAAF (A/401899) Pilot:
b 22 Apl 1911 Carlton, Australia.
Killed in Action 16 May 1943.

SERVICE	PROMOTIONS	AWARDS
Joined RAAF 1941		
Trainee 26 Apl 1941		
To Canada 18 Sep 1941	Plt Off 15 Jan 1942	
To England 2 Mar 1942		
No. 1654 CU		
61 Sqdn	Fg Off 16 Jul 1942	DFC Gaz
617 Sqdn Mar 1943	A/Fl Lt	14 May 1943 (29 ops)

Sergeant Samuel Leslie Whillis (144618) Flight Engineer:
b 1912 Newcastle-on-Tyne.
Killed in Action 16 May 1943.
Joined RAF 1940
Padgate 20 May 1940
No. 6 SOT 6 Sep 1940
152 Sqdn 31 Jan 1941
Warmwell 16 Mar 1941
No. 45 MU 7 7 Jul 1941
No. 13 MU 14 Jan 1942
No. 16 OTU 9 Jul 1942
No. 4 STT 18 Aug 1942
No. 1654 CU 1 Oct 1942
61 Sqdn
617 Sqdn 31 Mar 1943

Flying Officer Phillip Sidney Burgess (124881) Navigator:
b 1923
Killed in Action 16 May 1943.
Joined RAF 1941
No. 31 B & G Sch 25 May 1942 Plt Off 25 May 1942
No. 10 OTU 14 Jul 1942
No. 19 OTU 17 Aug 1942
No. 1661 CU 22 Nov 1942
61 Sqdn 17 Jan 1943   Fg Off 25 Nov 1942
617 Sqdn 31 Mar 1943

Flying Officer Charles Rowland Williams RAAF (A/405224) Wireless Operator:
b 1909 Torres Greek, Queensland, Australia.
Killed in Action 16 May 1943.

Joined RAAF 1941		
Trainee 3 Feb 1941		
To England 16 Oct 1941		
61 Sqdn	Plt Off 22 Sep 1941	DFC Gaz
		20 Jul 1945 (28 ops)
617 Sqdn Mar 1943	Fg Off 22 Mar 1942	wef 16 May 1943

Sergeant Alan Gillespie (1079863) Bomb aimer:
Carlisle.   Solicitor's clerk.
Killed in Action 16 May 1943.

SERVICE	PROMOTIONS	AWARDS
Joined RAF 1940		
Padgate 22 Nov 1940		
No. 7 RC 29 Nov 1940		
Wattisham 3 Jan 1941		
Andover Jan 1941		
No. 1 RW 18 Mar 1941		
Canada 15 Sep 1941		
No. 3 PRC 20 Feb 1942		
Wigton 23 Mar 1941		
No. 16 OTU 9 Jul 1942		
No. 1654 CU 30 Sep 1942		
61 Sqdn 27 Oct 1942		DFC Gaz
		20 Jul 1945 (33 ops)
617 Sqdn 31 Mar 1943	Plt Off 20 Mar 1943	wef 16 May 1943

Flying Officer Harvey Sterling Glinz RCAF (J/10212) Front Gunner:
b 1921 Winnipeg, Canada.
Killed in Action 16 May 1943.

Joined RCAF 1941		
Enlisted 5 Sep 1941	Commissioned	
61 Sqdn		
617 Sqdn 31 Mar 1943		

Sergeant Jack Robert George Liddell (1338282) Rear Gunner:
b 1925 Weston-super-Mare, Somerset.
Killed in Action 16 May 1943.

Joined RAF 1941		
Enlisted 30 May 1941		
Reserve 31 May 1941		
No. 3 RC 13 Oct 1941		
No. 10 (S) RC 15 Oct 1941		
Blackpool 28 Nov 1941		
No. 1 AAS 16 Dec 1941		
No. 7 AGS 16 May 1942		
No. 25 OTU 26 May 1942		
61 Sqdn 8 Sep 1942		
No. 1485 BG Fl 24 Mar 1943		
61 Sqdn 26 Mar 1943	Flew 30 operations	
617 Sqdn 31 Mar 1943		

Pilot Officer Geoffrey Rice (141707) Pilot:
b 4 Jan 1917.
Shot down 20 December 1943; taken prisoner six months later after hiding with Resistance.
Died November 1981, Somerset.

SERVICE	PROMOTIONS	AWARDS
Joined RAF 1941		
No. 2 RC 16 Jan 1941		
No. 8 ITW 14 Jun 1941		
Arnold 31 Jul 1941		
Canada 25 Aug 1941		
No. 31 PD 2 Nov 1941		
Maxwell AFB 9 Jan 1942	Plt Off 20 Feb 1942	
No. 31 PD 11 Mar 1942		
No. 3 PRC 29 Mar 1942		
No. 14 (P) AFU 9 Jun 1942		
No. 19 OTU 14 Jul 1942		
57 Sqdn 20 Feb 1943		
617 Sqdn 26 Mar 1943	Fl Off 20 Aug 1943	DFC Gaz
To UK 10 May 1945	Fl Lt 20 Jan 1945	16 Nov 1943
Pilot course 6 Aug 1945	Refresher	
21 AFU 13 Nov 1945		
13 OTU 23 Jul 1946		
21 Sqdn 15 Jan 1947	Germany	
Released 3 Jul 1947		

Sergeant Edward Clarence Smith (540655) Flight Engineer:
b 1919, Easthampstead, Berkshire.
Killed in Action 20 December 1943.

Joined RAF 1937
Uxbridge 11 May 1937
Hendon 12 Jun 1937
No. 3 STT 27 Aug 1937
19 Sqdn 26 Oct 1937
Hendon 1 Nov 1937
75 Sqdn 8 Jul 1938
No. 15 OTU 4 Apl 1940
Cosford 24 May 1940
32 Sqdn 21 Sep 1940
Ibsley 10 May 1941
130 Sqdn 25 Jun 1941
No. 1 Aug 1942
106 Sqdn 12 Oct 1942
57 Sqdn 9 Dec 1942
617 Sqdn 26 Mar 1943

Flying Officer Richard MacFarlane (126044) Navigator:
Killed in Action 20 December 1843.

SERVICE	PROMOTIONS	AWARDS
Joined RAFVR 1941		
No. 2 AFU 11 Jul 1942	Plt Off 11 Jul 1942	
No. 19 OTU 14 Jul 1942		
No. 1660 CU 23 Oct 1942		
57 Sqdn 9 Dec 1942		
617 Sqdn 26 Mar 1943		
Pilot's course 17 Oct 1943	Fg Off 17 Oct 1942	

Sergeant Chester Bruce Gowrie RCAF (R/93201) Wireless Operator:
b 15 April 1918 Saskatchewan, Canada.
Shot down 20 December 1943 – reported by the Red Cross to have been shot by the Germans.

Joined RCAF 1941
Ontario Mar 1941   Fl Sgt 22 Jun 1943
617 Sqdn Mar 1943

Flight Sergeant John William Thrasher RCAF (J/19337) Bomb aimer:
b 30 Jul 1920 Amherstburg, Ontario, Canada.
Killed in Action 20 December 1943.

Joined RCAF 1941
Enlisted 5 May 1941
617 Sqdn 26 Mar 1943

Sergeant Thomas William Maynard (1786420) Front Gunner:
b 1923 Wandsworth, S. London.
Killed in Action 20 December 1943.

Joined RAF 1941
Euston 12 Mar 1941
Reserve 13 Mar 1941
No. 3 RC 29 Dec 1941
No. 9 RC 2 Feb 1941
No. 15 RC 23 Jan 1942
No. 62 WU 24 Feb 1942
ACRE 6 Jul 1942
No. 14 ITW 1 Jul 1942
No. 1660 CU 20 Oct 1942
57 Sqdn 9 Dec 1942
1485 BG Flt 21 Mar 1943
617 Sqdn 26 Mar 1943

Sergeant Stephen Burns (1217692) Rear Gunner:
Dudley, Worcestershire.
Killed in Action 20 December 1943.

No. 2 RC 29 Jan 1941
No. 4 RC 7 Feb 1941
No. 13 OTU 21 Mar 1941

Sergeant Stephen Burns (1217692) Rear Gunner: (continued)

SERVICE	**PROMOTIONS**	**AWARDS**
No. 30 MU 3 Jun 1941		
No. 10 (S) RC 12 Aug 1941		
Blackpool 20 Oct 1941		
Church Fenton 8 Nov 1941		
No. 11 STT 17 Jun 1942		
57 Sqdn 21 Nov 1942		
617 Sqdn 26 Mar 1943		
Scampton 1 Aug 1943		
617 Sqdn 10 Aug 1943		
Scampton 22 Aug 1943		
Coningsby		
617 Sqdn 8 Sep 1943		
Coningsby 8 Oct 1943		
617 Sqdn Oct 1943		

Flight Lieutenant John Leslie Munro RNZAF (NZ/413942) Pilot:
b 5 Apl 1919 Giborne, New Zealand.   Farmer.
Lives in New Zealand.

Joined RNZAF 1941		
Enlisted 5 July 1941		
To UK 20 Oct 1941		
97 Sqdn 1942/43		DFC Gaz
		11 Jun 1943 (21 ops)
617 Sqdn 25 Mar 1943		DSO Gaz 28
No. 1690 BDTF 13 Jul 1944	A/Sdn Ldr 14 Feb 194	28 Apl 1944 (41 ops)
Released 5 Feb 1946		

Sergeant Frank Ernest Appleby (652494) Flight Engineer:

Joined RAF 1939		
No. 3 RC 10 Aug 1939		
No. 3 Wing 22 Sep 1939		
Martlesham		
No. 7 STT		
No. 45 MU 18 Jul 1941		
No. 1654 CU 1 Oct 1942		
97 Sqdn Dec 1942	T/F/Sgt 1 Mar 1943	
617 Sqdn Mar 1943		DFM Gaz
No. 5 LFS 26 Jul 1943		30 Jun 1944 (39 ops)
No. 5 PDC 14 Sep 1944		
No. 45 Group 17 Sep 1944		
No. 11 PDC 7 Jan 1946		

Flying Officer Francis Grant Rumbles (120350) Navigator:
Living South Africa.
Joined RAFVR 1940
Enlisted 10 Oct 1940
No. 42 Air Sch 4 Apl 1942

Flying Officer Francis Grant Rumbles (120350) Navigator: (continued)

SERVICE	PROMOTIONS	AWARDS
No. 3 AFU 8 Jun 1942		
No. 29 OTU 17 Jul 1942		
No. 1654 CU 30 Sep 1942		
97 Sqdn 11 Dec 1942	Pl Off 4 April 1942	DFC Gaz
		15 Jun 1943 (23 ops)
617 Sqdn 25 Mar 1943	Fg Off 4 Oct 1942	Bar DFC
Waddington 18 Aug 1944	Flew with 9 Sqdn,	15 Jun 1944 (56 ops)
HQ 118 Wing 23 Oct 1945	Nav. Leader. (Tirpitz Op)	
Japan 15 Feb 1946	Fl Lt 4 April 1944	
Released 14 Feb 1947	A/Sqn Ldr	

Sergeant Percy Edgar Pigeon RCAF (J/97620) Wireless Operator:
b June 1917 Williams Lake, British Columbia, Canada.
Retired from RCAF 6 December 1962 as Wing Commander.
Died 23 March 1967.

97 Sqdn 1942		
617 Sqdn Mar 1943	F/Sgt 1 Sep 1943	DFC Gaz
No. 86 OTU 12 Jul 1944	Plt Off 7 Dec 1943	30 Jun 1944 (46 ops)
To Canada Dec 1944		

Sergeant James Henry Clay (159893) Bomb aimer:
Living Newcastle-upon-Tyne.

97 Sqdn Oct 1942	T/F/Sgt 7 Sept 1943.	
	One tour	
617 Sqdn Mar 1943	Plt Off 17 Nov 1943	DFC Gaz
14 OTU 14 Jun 1945		30 Jun 1944 (46 ops)
No. 26 OTU 1945		
No. 1656 CU 17 Jul 1945		
Retired 3 Dec 1945		

Sergeant William Howarth (1479639) Front Gunner:
Oldham Lancs.
Lives in Oldham, Lancashire.

Joined RAF 1941		
No. 3 RC 5 Jun 1941		
No. 10 (S) RC 20 Oct 1941		
Blackpool 22 Dec 1941		
Gravesend 29 Jan 1942		
No. 11 STT 17 Jun 1942		
No. 7 AGS 22 Aug 1942		
97 Sqdn 11 Dec 1942		
617 Sqdn 25 Mar 1943	Plt Off 4 Jun 1944	DFM Gaz
No. 10 OTU	Fg Off 4 Oct 1944	30 Jun 1944 (41 ops)
	Fl Lt 4 April 1945	
93 & 91 Gps	Courses	
CGS	Gunnery Leaders	
Finningley	Course.	
Silverston	Gunnery Leader	
Released 6 Jun 1946		

Flight Sergeant Harvey Alexander Weeks RCAF (J/19206) Rear Gunner:
Living in Canada.

SERVICE	PROMOTIONS	AWARDS
617 Sqdn	Plt Off 1944	DFC Gaz
BDTF 28 Jul 1944		30 Jun 1944 (46 ops)

Flight Sergeant William Clifford Townsend (656738) Pilot:
b 12 Jan 1921 Brookend, Gloucester.
Lives in the Midlands.

Joined Army 1941		
Royal Artillery OCTU		
12 Jan 1941		
To RAF 14 May 1941	AC2	
No. 9 RW 24 May 1941		
No. 11 ITW 31 May 1941		
No. 51 Gp Pool 19 Jul 1941		
No. 11 FTS 6 Sep 1941		
No. 14 FTS Sep 1941		
No. 12 FTS 1 Jan 1942	Sgt 24 Jan 1942	
BAT Flt Mar 1942		
No. 16 OTU 10 Mar 1942	F/Sgt 1 Apl 1942	
49 Sqdn 12 Jun 1942	Flew 26 operations	DFM Gaz
	Plt Off 16 Mar 1943	14 May 1943
617 Sqdn 25 Mar 1943	Fg Off 19 Sep 1943	CGM Gaz
No. 1668 CU 17 Sep 1943		28 May 1943
No. 29 OTU 31 Oct 1943		
No. 85 OTU Jul 1944		
No. 7 FIS Jul 1944		
India Feb 1945	Fl Lt 16 Mar 1945	
No. 2 PDC 21 Mar 1945		
HQ ACSEA 25 Mar 1945		
HQ 226 Gp Apl 1945		
No. 6 RFU May 1945		
ACSEA 7 Jul 1945		
HQ 229 GP 28 Nov 1945		
No. 104 PDC 12 Apl 1946		
Released 3 May 1946		

Sergeant Dennis John Dean Powell (644741) Flight Engineer:
b 1922 Sidcup, Kent.
Killed in Action 16 September 1943.

Joined RAF 1939		
Linton/Ouse 22 May 1939		
Cosford 12 Aug 1939		
No. 10 B & G Sch 19 Jan 1940		
No. 2 STT 8 Nov 1940		
No. 1 AAS 9 Jan 1941		
No. 4 STT 28 Jul 1941		
49 Sqdn 12 Oct 1942		
617 Sqdn 26 Mar 1943		MinD Gaz
		7 Jun 1943

Pilot Officer Cecil Lancelot Howard RAAF (A/406248) Navigator:
b 12 Jan 1913 Sth Freemantle, W. Australia.
Lives in Western Australia.

SERVICE	PROMOTIONS	AWARDS
Joined RAAF 1940		
Trainee 7 Oct 1940		
To England 16 Oct 1941	Plt Off 21 Jan 1943	
49 Sqdn 1942	Flew 25 operations	
617 Sqdn 25 Mar 1943	Fg Off 21 Jul 1943	DFC Gaz
No. 1654 CU 6 Oct 1943		28 May 1943
Australia 12 May 1944	Fl Lt 21 Jul 1944	
Released 19 Mar 1945		

Flight Sergeant George Alexander Chalmers (552201) Wireless Operator:
Harrogate, Yorks.
Lives in Yorkshire.

Joined RAF 1938		
West Drayton 14 Jan 1938		
No. 3 EWS 26 Mar 1938		
10 Sqdn 18 May 1939		
7 Sqdn 25 Aug 1940		
35 Sqdn 18 Nov 1940		
No. 1652 CU 1 Feb 1942		
No. 1501 BAT F1 4 Mar 1942		
617 Sqdn 6 Apl 1943	Plt Off 26 Jun 1943	DFM Gaz
		28 May 1943 (44 ops)
No. 52 Base 12 Jul 1944	Fg Off 27 Dec 1943	DFC Gaz
Retired 12 Aug 1954	Fl Lt 27 Jun 1944	13 Oct 1944 (65 ops)

Sergeant Charles Ernest Franklin (1165320) Bomb aimer:
b 1913 Radlett, Herts.   Car worker.
Died 25 January 1975, Birmingham.

Joined RAF 1940		
No. 2 RC 18 Jun 1940		
Reserve 19 Jun 1940		
No. 10 (S) RC 28 Aug 1940		
PACE 30 Nov 1940		
Watton 22 Nov 1940		
No. 9 AGS 15 Sep 1941		
No. 19 OTU		
No. 25 OTU 21 Nov 1941		
44 Sqdn 6 Apl 1942		
49 Sqdn 21 Apl 1942	Flew 28 operations	DFM Gaz
		18 May 1943 (26 ops)
617 Sqdn 25 Mar 1943		Bar DFM
No. 1660 CU 30 Aug 1943		28 May 1943
No. 1 AAS 8 Jan 1944		
83 Sqdn 4 May 1944	Plt Off 4 May 1944	
	Fg Off 4 Nov 1944	

Sergeant Douglas Edward Webb (1517241) Front Gunner:
b 1922 Leytonstone, Essex.
Lives in Surrey.

SERVICE	PROMOTIONS	AWARDS
Joined RAF 1940		
No. 2 RC 10 Dec 1940		
No. 7 RC		
No. 15 OTU 14 Feb 1941		
No. 10 (S) RC 20 Jul 1941		
No. 2 Sig Sch 31 Oct 1941		
109 Sqdn Mar 1942		
ACRC 5 May 1942		
No. 14 ITW 22 May 1942		
49 Sqdn 15 Mar 1943	Flew 25 operations	
617 Sqdn 25 Mar 1943		DFM Gaz
No. 1667 CU 2 Oct 1943		28 May 1943
No. 1661 CU 6 Oct 1943		
No. 5 LFS 21 Nov 1943		
617 Sqdn 19 Oct 1944		

Sergeant Raymond Wilkinson (1511241) Rear Gunner:
b 1922 South Shields, Co. Durham.   Joiners' App.
Died in 1980.

	PROMOTIONS	AWARDS
Joined RAF 1941		
No. 3 RC 4 Oct 1941		
No. 9 RC 20 Dec 1941		
Coltishall 13 Feb 1942		
ACRC 22 Jun 1942		
No. 14 ITW 27 Jun 1942		
No. 9 AGS 18 Jul 1942		
No. 1654 CU 24 Aug 1942		
49 Sqdn 20 Sep 1942	Flew 21 operations	
No. 1485 BGF 16 Mar 1943		
49 Sqdn 18 Mar 1943		
617 Sqdn 26 Mar 1943		DFM Gaz
No. 1662 CU 2 Oct 1943		28 May 1943
No. 1668 CU		
No. 5 LFS 21 Nov 1943	Plt Off 9 Feb 1944	
617 Sqdn 19 Oct 1944	Fg Off 10 Aug 1944	
Released 18 Sep 1946	Fl Lt 10 Feb 1946	

Flight Sergeant Cyril Thorpe Anderson (518252) Pilot:
b Wakefield, Yorkshire.   Engineering apprentice.
Killed in Action 23 September 1943 on 23rd trip.

Joined RAF 1934
Trg Dept 22 Oct 1934
Gosport 19 Feb 1935
Singapore 1 Apl 1936
Donisbristle 15 Nov 1937

Flight Sergeant Cyril Thorpe Anderson (518252) Pilot: (continued)

SERVICE	PROMOTIONS	AWARDS
Thorney Is 25 Mar 1938		
No. 1 RW 3 May 1941		
No. 3 ITW 17 May 1941		
51 Gp Pool 12 Jul 1941		
No. 33 SFT 9 Sep 1941	Canada	
No 7 Depot 3 Jan 1942		
RAF TP 24 Jan 1942		
No. 3 PRC 10 Feb 1942		
No. 6 SFTS 1 Apl 1942		
No. 25 OTU 7 Jun 1942		
No. 1654 CU 12 Jan 1943		
49 Sqdn 23 Feb 1943	Flew 9 operations	
617 Sqdn 24 Mar 1943	Flew 1 operation	
49 Sqdn 2 Jun 1943	Flew 13 operations	
	Plt Off 17 Apl 1943	

Sergeant Robert Campbell Patterson (628327) Flight Engineer:
b 1907 Edinburgh.
Killed in Action 23 September 1943.
    Joined RAF 1938
    Upwood 1 Dec 1938
    Henlow 10 Feb 1939
    Locking 21 Mar 1939
    144 Sqdn 3 Nov 1939
    No. 2 STT 27 Sep 1940
    No. 13 MU 25 Nov 1940
    No. 53 OTU 8 Jul 1941
    RNAS Lee 28 May 1942
    RNAS Manston 3 Jun 1942
    No. 4 STT 24 Jun 1942
    No. 1654 CU 12 Jan 1943
    49 Sqdn 23 Feb 1943
    617 Sqdn 24 Mar 1943
    49 Sqdn 6 Jun 1943

Sergeant John Percival Nugent (1265662) Navigator:
b 9 Aug 1914 Stony Middleton, Derbyshire.   Teacher.
Killed in Action 23 September 1943.
    Joined RAF 1940
    Uxbridge 14 Sep 1940
    No. 9 RC 16 Sep 1940
    Drem 25 Oct 1940
    No. 10 (S) RC 26 Jan 1941
    No. 8 SFTS 12 Apl 1941
    No. 14 OTU 11 Jun 1941
    No. 1 ACRC 26 Sep 1941
    No. 31 Depot 25 Nov 1941

Sergeant John Percival Nugent (1265662).Navigator: (continued)

SERVICE	PROMOTIONS	AWARDS
Trenton 13 Jan 1942		
PTC	Toronto	
Moncton 29 Jan 1942		
No. 31 PD 14 May 1942		
No. 3 PRC 12 Jun 1942		
No. 3 (O) AFU 27 Sep 1942		
No. 25 OTU 8 Oct 1942		
No. 1654 CU 12 Jan 1943		
49 Sqdn 24 Feb 1943		
617 Sqdn 24 Mar 1943		
49 Sqdn 2 Jun 1943		

Sergeant William Douglas Bickle (1380667) Wireless Operator:
Killed in Action 23 September 1943.

    Joined RAF 1940
    No. 2 RC 11 Oct 1940
    No. 7 RC 18 Oct 1940
    No. 17 OTU 20 Nov 1940
    No. 10 (S) RC 11 Apl 1941
    No. 2 Sig Sch 29 Aug 1941
    No. 1 AACU 29 Nov 1941
    No. 2 Sig Sch 24 Jun 1942
    No. 3 AGS 8 Nov 1942
    No. 25 OTU 8 Sep 1942
    No. 1654 CU 12 Jan 1943
    49 Sqdn 23 Feb 1943
    617 Sqdn 29 Mar 1943
    49 Sqdn 6 Jun 1943

Sergeant Gilbert John Green (1322942) Bomb aimer:
b 1922 Southall, Middlesex.
Killed in Action 23 September 1943.

    Joined RAF 1941
    Oxford 30 Jul 1941
    Reserve 31 Jul 1941
    No. 1 ACRC 27 Oct 1941
    ACDW 13 Dec 1941
    No. 6 ITW 3 Jan 1942
    51 Gp Pool 29 Jan 1942
    ACDC 10 Jun 1942
    No. 2 (O) AFU 13 Jun 1942
    No. 25 OTU 8 Sep 1942
    No. 1654 CU 12 Jan 1943
    49 Sqdn 23 Feb 1943
    617 Sqdn 24 Mar 1943
    49 Sqdn 6 Jun 1943

Sergeant Eric Ewan (1576002) Front Gunner:
b 1922.
Killed in Action 23 September 1943.

SERVICE	PROMOTIONS	AWARDS
Joined RAF 1941		
Birmingham 28 Jun 1941		
Reserve 29 Jun 1941		
No. 3 RC 30 Oct 1941		
No. 10 RC 3 Nov 1941		
RDU 8 Dec 1941		
No. 9 RC 30 Dec 1941		
No. 15 RC 23 Jan 1942		
No. 57 DTS 9 Feb 1942		
ACRC 15 Jun 1942		
No. 14 ITW 27 Jun 1942		
No. 2 AGS 24 Jul 1942		
No. 1654 CU 12 Jan 1943		
617 Sqdn 24 Mar 1943		
49 Sqdn 2 Jun 1943		

Sergeant Arthur William Buck (1305690) Rear Gunner:
London.
Killed in Action 23 September 1943.

Joined RAF 1940		
Blackpool 22 Jul 1940		
No. 93 MU 7 Aug 1940		
Abbey Lodge 24 Sep 1942		
No. 14 ITW 2 Oct 1942		
No. 7 AGS 31 Oct 1942		
No. 1654 CU 10 Jan 1943		
49 Sqdn 23 Feb 1943		
617 Sqdn 24 Mar 1943		
49 Sqdn 6 Jun 1943		

Flight Sergeant Kenneth William Brown RCAF (R/94567) Pilot:
b Aug 1920 Moose Jaw, Saskatchewan, Canada.
Living in Canada.

Joined RCAF 1941		
PRC Feb 1942		
Waterbeach 1 May 1942		
No. 1651 CU		
No. 6 AFU 23 Jun 1942		
No. 19 OTU 11 Aug 1942		
No. 1660 CU 15 Dec 1942		
No. 1654 CU Dec 1942		
44 Sqdn 5 Feb 1943	Flew 7 operations.	
617 Sqdn Mar 1943	Plt Off 20 Apl 1943	CGM Gaz
No. 5 LFS 3 May 1944	Instructor	28 May 1943
No. 6 Gp	Fl Lt 7 Oct 1943	
Pacific		
Canada 1945		
Retired RCAF 1967	Sqdn Ldr rank	

Sergeant Harry Basil Feneron (1266419) Flight Engineer:
Living in Buckinghamshire.

SERVICE	PROMOTIONS	AWARDS
Joined RAF 1940		
Penarth 17 Sep 1940	AC2	
No. 9 RC 31 Oct 1940		
No. 10 STT 6 Dec 1940		
No. 7 STT 11 Jul 1941		
Hendon 26 Sep 1941		
No. 4 STT 23 Sep 1942		
No. 1654 CU 21 Oct 1942		
44 Sqdn 5 Feb 1943		
617 Sqdn 25 Mar 1943	Plt Off 13 Jan 1944	
No. 1654 CU 28 Mar 1944	Fg Off 13 Jul 1944	
	Fl Lt 13 Jan 1946	
75 Base Gp 26 Apl 1945		
61 Base 11 Jun 1945		
Syserston 9 Jul 1945		
Upwood 16 Oct 1945		
115 Sqdn 5 Jan 1946		
Released 12 Apl 1946		

Sergeant Dudley Percy Heal (919764) Navigator:
b 1916 Hampshire.

Joined RAF 1940		
Uxbridge 29 Mar 1940		
Reserve 30 Mar 1940		
Uxbridge 8 May 1940		
No. 4 RC 14 May 1940		
Wyton 1 Jun 1940		
Southampton 20 Jun 1940		
No. 53 OTU 10 Mar 1941		
No. 1 RW 10 May 1941		
No. 8 ITW 17 May 1941		
PTC 19 Jul 1941	Toronto	
No. 31 PD 18 Nov 1941		
Moncton		
No. 31 PD 9 May 1942		
No. 3 PDC 18 May 1942		
No. 3 (O) AFU 20 Jun 1942		
No. 19 OTU 18 Aug 1942		
434 Sqdn 22 Dec 1942		
44 Sqdn 4 Feb 1943	Flew 6 operations	
617 Sqdn 25 Mar 1943		DFM Gaz
No. 29 OTU 26 Mar 1944		18 May 1943
214 Sqdn 8 Feb 1945		
No. 2 Embarkation Unit		
Southampton 11 Sept 1945		
Released 11 Mar 1946		

Sergeant Harry J. Hewstone (1378012) Wireless Operator:
Living in Essex.

SERVICE	PROMOTIONS	AWARDS
Joined RAF 1939		
Blackpool 14 Sep 1939		
Scampton 23 Oct 1940		
No. 10 (S) RC 30 Jan 1941		
54 Sqdn 12 Sep 1941		
No. 2 Sig Sch 13 May 1942		
No. 1 AGS 27 Jun 1942		
No. 19 OTU 31 Aug 1942		
617 Sqdn 26 Mar 1943		

Sergeant Stefan Oancia RCAF (R/114949) Bomb aimer:
b 1923 Stonehenge, Saskatchewan, Canada.
Now lives in the USA.

Joined RCAF 1941		
Enlisted 1 Aug 1941		
ITU 4 Dec 1941		
BLG 25 Apl 1942		
AOS 13 Mar 1942		
ANS 25 May 1942		
617 Sqdn		DFM Gaz
No. 11 OTU 26 Mar 1944		28 May 1943 (6 ops)

Sergeant Daniel Allatson (553739) Front Gunner:
b 1924 Upminster, Essex.
Killed in Action 16 September 1943.

Joined RAF 1939		
Uxbridge 15 Feb 1939		
No. 2 AAS 18 Feb 1939		
Cosford 3 Sep 1939		
Hucknall 12 Dec 1939		
No. 18 OTU 8 May 1940		
No. 1 (P) FTS 28 Nov 1940		
ACRC 16 Feb 1942		
No. 4 ITW 7 Mar 1942		
ACDW 30 Mar 1942		
Marston 10 Apl 1942		
279 Sqdn 4 May 1942		
No. 11 STT 9 Sep 1942		
No. 1660 CU 6 Jan 1943		
57 Sqdn 17 Feb 1943		
617 Sqdn 10 Apl	In Divall's crew–	
	replacement for raid	

Flight Sergeant Grant S. MacDonald Rear Gunner:

44 Sqdn		
617 Sqdn		
No. 29 OTU 26 Mar 1944		

Pilot Officer Warner Ottley (141460) Pilot:
Killed in Action 16 May 1943.

SERVICE	PROMOTIONS	AWARDS
Joined RAF 1941		
No. 1 RC 20 Jan 1941	No. 1332172	
No. 54 Gp		
No. 8 ITW 8 Mar 1941		
No. 31 EFTS 15 Jun 1941		
No. 34 SFTS 12 Aug 1941		
'Y' Depot 24 Oct 1941		
No. 3 PRC 14 Nov 1941		
No. 3 SFTS 2 Dec 1941		
No. 3 PRC 16 Jan 1942		
No. 19 OTU 19 Jan 1942		
50 Sqdn 18 Jun 1942		
83 Sqdn 19 Jun 1942		
No. 207 CF 5 Oct 1942		
207 Sqdn 12 Nov 1942	Flew 31 operations	DFC Gaz
	Plt Off 25 Jan 1943	29 Jun 1945
617 Sqdn 17 Mar 1943		wef 16 May 1943

Sergeant Ronald Marsden (568415) Flight Engineer:
b 1920 Redcar, Yorkshire.   Boy Entrant.
Killed in Action 16 May 1943.
Joined RAF 1935
Halton 27 Aug 1935
Abingdon 18 Aug 1938
Thornaby 1 Sep 1938
No. 14 OTU 5 Apl 1940
Sealand 29 Jun 1941
No. 14 OTU 4 Aug 1941
No. 10 AGS 13 Jun 1942
No. 14 STT 29 Jul 1942
No. 1654 CU 18 Sep 1942
61 Sqdn 7 Oct 1942
No. 1660 CU 20 Oct 1942
617 Sqdn 6 Apl 1943

Flying Officer Jack Kenneth Barrett (15775) Navigator:
b 1921 Goodmayes, Essex.
Killed in Action 16 May 1943.
Joined RAF 1940
No. 1 RC 9 May 1940
Reserve 10 May 1940
No. 1 RW 20 Sep 1940
No. 2 TW 5 Oct 1940
Sth Africa 6 Jan 1941
No. 45 Air Sch 10 Feb 1941   Plt Off 13 Jul 1941

Flying Officer Jack Kenneth Barrett (115775) Navigator: (continued)

SERVICE	PROMOTIONS	AWARDS
To UK 15 Jul 1941		
No. 3 PRC 23 Aug 1941		
No. 25 OTU 16 Sep 1941		
207 Sqdn 28 Feb 1942	Fg Off 13 Jul 1942	
No. 1660 CU 20 Oct 1942		
207 Sqdn 12 Nov 1942		DFC Gaz
		29 Jun 1945
617 Sqdn 6 Apl 1943		wef 16 May 1943

Sergeant Jack Guterman (1172550) Wireless Operator:
b 1920 Guildford, Surrey.
Killed in Action 16 May 1943.

Joined RAF 1940		
Cardington 16 Jul 1940		
Reserve 17 Jul 1940		
Blackpool 23 Aug 1940		
No. 2 SST 22 Nov 1940		
No. 2 SAC 1 Feb 1941		
No. 2 Sig Sch 25 Feb 1941		
No. 5 AOS 22 Aug 1941		
No. 23 OTU 16 Sep 1941		
207 Sqdn 25 Feb 1942		
No. 1660 CU 20 Oct 1942		
207 Sqdn 12 Nov 1942	Flew 28 operations	DFM Gaz
617 Sqdn 6 Apl 1943		14 May 1943

Flight Sergeant Thomas Bar Johnston (1060657) Bomb aimer:
Bellshill, Lanarkshire.
Killed in Action 16 May 1943.

Joined RAF 1940
Padgate 22 July 1940
Reserve 23 Jul 1940
No. 1 RW 11 Jan 1941
No. 8 ITW 8 Feb 1941
No. 33 ANS 22 Jun 1941
No. 31 B & G S 26 Oct 1941
No. 16 AFTB 7 Jan 1942
No. 3 PRC 20 Jan 1942
No. 2 AOS 12 Feb 1942
207 Sqdn 31 Aug 1942
No. 207 CU 7 Sep 1942
No. 1660 CU 20 Oct 1942
207 Sqdn 12 Nov 1942
617 Sqdn 6 Apl 1943

Flight Sergeant Frank Tees (1332270) Rear Gunner:
b 1923   Shot down 16 May 1943.   Died March 1982
Badly burned and taken prisoner, eventually going to POW Camp L6 at Heydekruge.

SERVICE	PROMOTIONS	AWARDS
Joined RAF 1941		
Uxbridge 6 Jan 1941		
No. 1 RC 14 Jan 1941		
No. 7 RC		
No. 3 SGR 14 Mar 1941		
No. 10 SRC 20 Aug 1941		
Hatfield Nov 1941		
No. 50 GRD 22 Nov 1941		
ACRC 2 Jun 1942		
No. 14 ITW		
No. 2 AGS 17 Jul 1942		
AGS Dunholm 23 Sep 1942		
CU Swinderby 7 Oct 1942		
207 Sqdn 12 Nov 1942		
617 Sqdn 6 Apl 1943		
No. 106 PRC 6 May 1945		
No. 225 MU 10 Sep 1945		
No. 106 PRC 30 Dec 1945		
No. 104 PDC 5 Jan 1946		

Sergeant Harry John Strange (1395453) Front Gunner:
b 1923 Holloway, London.
Killed in Action 16 May 1943.
    Joined RAF 1941
    Euston 26 Aug 1941
    Reserve 27 Aug 1941
    No. 3 RC 4 Dec 1941
    No. 9 RC 9 Dec 1941
    No. 15 RC 23 Jan 1942
    Ford 5 Feb 1942
    ACSB
    ACRC
    No. 14 ITW 4 Jul 1942
    No. 4 AGS 28 Aug 1942
    No. 1660 CU 7 Nov 1942
    207 Sqdn 9 Nov 1942
    617 Sqdn 6 Apl 1943

Pilot Officer Lewis Johnstone Burpee RCAF (J/17115) Pilot:
b. 5 Mar 1918 Ottawa, Ontario, Canada.   BA Degree.
Killed in Action 17 May 1943.

Joined RCAF 1940		
Enlisted 17 Dec 1940		
To England Sep 1941		
106 Sqdn 1942		DFM Gaz
617 Sqdn 29 Mar 1943	Plt Off 5 Mar 1943	18 May 1942 (26 ops)

Sergeant Guy Pegler (573474) Flight Engineer:
b 1921 Bath, Somerset.
Killed in Action 17 May 1943.

SERVICE	PROMOTIONS	AWARDS
No. 1 STT 25 Jan 1938		
107 Sqdn 2 Jul 1940		
124 Sqdn 31 May 1941		
154 Sqdn 24 Nov 1941		
No. 4 STT 28 Jul 1942		
No. 106 CU 30 Aug 1942		
106 Sqdn 10 Oct 1942		
617 Sqdn 29 Mar 1943		

Sergeant Thomas Jaye (1299446) Navigator:
b 3 Oct 1922 Crook, Co. Durham.   Elec Engineer.
Killed in Action 17 May 1943.

Joined RAF 1941
No. 1 RC 3 Mar 1941
No. 7 RC 4 Mar 1941
Kenley 18 Apl 1941
No. 1 ITW 24 Jun 1941
Bircham Newton 28 Jun 1941
PAA 14 Oct 1941
No. 31 PD 25 Oct 1941
No. 3 PRC 8 Apl 1942
No. 25 OTU 14 Jul 1942
106 Sqdn 26 Dec 1942
617 Squadron 29 Mar 1943

Pilot Officer Leonard George Weller (142507) Wireless Operator:
b 1915 Harrow, Middlesex.
Killed in Action 17 May 1943.

Joined RAF 1940
No. 9 RC 22 Aug 1940
96 Sqdn 11 Sep 1940
No. 51 Gp Pool 26 Sep 1940
No. 1 RW 11 Oct 1940
No. 4 ITW 19 Oct 1940
No. 51 Gp Pool 8 Jan 1941
No. 10 (S) RC 29 Apl 1941
No. 2 Sig Sch 22 Aug 1941
Mount Batten 21 Nov 1941
172 Sqdn 28 Apl 1942
No. 2 Sig Sch 31 Jul 1942
No. 8 AGS 31 Jul 1942
No. 16 OTU 1 Sep 1942
No. 1654 CU 12 Dec 1942
106 Sqdn 5 Feb 1943   Plt Off 21 Feb 1943
617 Sqdn 29 Mar 1943

Sergeant James Lamb Arthur RCAF (R/119416) Bomb aimer:
b 1918 Coldwater, Ontario, Canada.
Killed in Action 17 May 1943.

SERVICE	PROMOTIONS	AWARDS
Joined RCAF 1941		
Enlisted 29 Jul 1941		
617 Sqdn 29 Mar 1943		

Sergeant William Charles Arthur Long (1600540) Front Gunner:
b 1924 Bournemouth, Hampshire.
Killed in Action 17 May 1943.
Joined RAF 1941
Oxford 8 Oct 1941
Reserve 9 Oct 1941
No. 1 ACRC 23 Mar 1942
No. 11 ITW 11 Apl 1942
ACDW 22 May 1942
No. 14 ITW 23 May 1942
No. 4 AGS 28 Jun 1942
Coningsby 10 Aug 1942
106 Sqdn 28 Sep 1942
617 Sqdn 29 Mar 1943

Flight Sergeant Joseph Gordon Brady RCAF (R/93554) Rear Gunner:
b 1916 Ponoka, Alberta, Canada.
Killed in Action 17 May 1943.
Joined RCAF 1941
Enlisted 15 Mar 1941
617 Sqdn 29 Mar 1943

The following aircrew members trained for the Raid but due to illness in the crews did not take part in the attack.

Flight Lieutenant Harold Sydney Wilson (118566) Pilot:
b 1915 Tottenham, Middlesex.
Killed in Action 16 September 1943.
Joined RAF 1940
Uxbridge 27 Nov 1940
No. 1 RW 17 May 1941
No. 3 ITW 31 May 1941
No. 51 Gp Pool 12 Jul 1941
Cranwell 25 Oct 1941
No. 3 SFTS
57 Sqdn
617 Sqdn Apl 1943

Sergeant Thomas William Johnson (612035) Flight Engineer:
Killed in Action 16 September 1943.

SERVICE	PROMOTIONS	AWARDS
Joined RAF 1938		
Uxbridge 10 May 1938	Plt Off 27 Mar 1942	
No. 3 STT 22 Jul 1938	Fg Off 1 Oct 1942	
Henlow 21 Sep 1938	A/Fl/Lt 24 Mar 1943	
St Athan 12 Oct 1938		
Thornaby 27 Jul 1939		
Cottesmore 25 Aug 1939		
No. 14 OTU 5 Apl 1940		
No. 7 AAMU 13 Sep 1940		
Cosford 28 Sep 1940		
No. 6 STT 26 Oct 1940		
No. 81 OTU 13 Jul 1941		
No. 1654 CU 1 Oct 1942		
44 Sqdn 22 Dec 1943		
617 Sqdn 25 Mar 1943		

Sergeant Clifford Morrell Knox (1029237) Navigator:
Killed in Action 16 September 1943.
Joined RAF 1941
No. 3 RC 6 Oct 1941
No. 10 (S) RC 8 Oct 1941
RDU 12 Jan 1942
York 3 Feb 1942
No. 11 STT 17 Jun 1942
No. 7 AGS 22 Aug 1942
No. 44 CF 5 Oct 1942
No. 1485 Gt Flt 11 Oct 1942
No. 1661 CU 9 Nov 1942
44 Sqdn Dec 1942
617 Sqdn Mar 1943

Sergeant Lloyd Mieyette RCAF Wireless Operator:
Killed in Action 16 September 1943.
Joined RCAF 1941
Enlisted 6 Jun 1941
617 Sqdn 25 Mar 1943   Warrant Officer

Pilot Officer George Henry Coles RCAF (J/16476) Bomb aimer:
Killed in Action 16 September 1943.
Joined RCAF 1941
Enlisted 28 Feb 1941
617 Sqdn 25 Mar 1943

Sergeant Trevor Herrington Payne (1257191) Front Gunner:
b 1920 Gateshead, Co. Durham.
Killed in Action 16 September 1943.

SERVICE	PROMOTIONS	AWARDS
Joined RAF 1940		
Uxbridge 13 Jul 1940		
No. 9 RC 14 Aug 1940		
Cranwell 3 Sep 1940		
PRU Heston 9 Oct 1940		
Scampton 20 Jan 1941		
No. 724 (D) Sqdn 11 Aug 1941		
No. 819 (D) Sqn 4 Sep 1941		
No. 14 ITW 25 Oct 1941		
No. 1 AAS 12 Dec 1941		
No. 9 AGS 13 Dec 1941		
No. 19 OTU 16 Apl 1942		
44 Sqdn 31 Aug 1942		
No. 44 CF 8 Nov 1942		
44 Sqdn 1 Jan 1943		
617 Sqdn 25 Mar 1943		

Sergeant Eric Hornby (1304938) Rear Gunner:
Killed in Action 16 September 1943.

Joined RAF 1940
No. 9 RC 18 Jul 1940
Wittering 2 Aug 1940
N. Luffenham 5 Feb 1941
No. 797 (D) Sqn 11 Aug 1941
ACRC 14 Nov 1941
No. 14 ITW 22 Nov 1941
No. 4 AGS 23 May 1942
No. 19 OTU 14 Jul 1942
No. 1654 CU 1 Oct 1942
44 Sqdn 22 Dec 1942
617 Sqdn 25 Mar 1943

Sergeant William George Divall (1387753) Pilot:
b 1922 Thornton Heath, Surrey.
Killed in Action 16 September 1943.

Joined RAF 1941		
Euston 4 Apl 1941		
No. 1 ACRC 28 Jul 1941		
No. 1 ITW 9 Aug 1941		
ATTS 14 Oct 1941		
No. 31 PD 29 Oct 1941	Canada	
No. 3 PRC 25 Jun 1942		
No. 19 OTU 25 Aug 1942		
No. 10 OTU 16 Jan 1943		
No. 1660 CU 5 Feb 1943		
57 Sqdn 17 Feb 1943	Plt Off 6 Mar 1943	
617 Sqdn 10 Apl 1943	Fg Off 6 Sep 1943	

Sergeant Ernest Cecil Allan Blake (548871) Flight Engineer:
b 1919 Horton, Bradford, Yorks.
Killed in Action 16 September 1943.

SERVICE	PROMOTIONS	AWARDS
Joined RAF 1938		
Uxbridge 10 Feb 1938		
No. 3 STT 29 Apl 1938		
Henlow 29 Jun 1938		
St Athan 12 Oct 1938		
32 Sqdn 27 Jul 1939		
Cosford 20 May 1940		
No. 19 OTU 6 Sep 1940		
St Athan 23 Mar 1941		
No. 32 MU 23 Feb 1942		
No. 24 MU 25 Mar 1942		
No. 1654 CU 12 Dec 1942		
57 Sqdn 17 Feb 1943		
617 Sqdn 10 Apl 1943		

Flying Officer Douglas William Warwick RCAF (J/11112) Navigator:
Toronto, Canada.
Killed in Action 16 September 1943.
Joined RCAF 1941
Enlisted 20 May 1941
617 Sqdn 11 Apl 1943

Sergeant James Stevenson Simpson (630257) Wireless Operator:
b 8 Sep 1920 Bo'ness, Scotland.   Apprentice slater platerer.
Killed in Action 16 September 1943.
Joined RAF 1939
Cardington 5 Jan 1939
No. 2 E & EWS 24 Mar 1939
9 Sqdn 10 Sep 1939
Farnborough 29 Sep 1939
HQ CFF 7 Oct 1939
No. 50 Wing 10 Oct 1939
Loaned to Army
France
Dunkirk May 1940
613 Sqdn 10 Oct 1940
No. 2 Sig Sch 9 Jan 1942
No. 4 Air Sch 16 Apl 1942
No. 8 AGS 1 May 1942
No. 19 OTU 16 Jun 1942
No. 10 OTU 16 Nov 1942
No. 1660 CU 5 Jan 1943
57 Sqdn 7 Feb 1943
617 Sqdn 10 Apl 1943

Sergeant Robert Campbell McArthur (1551189) Bomb aimer:
b 1922 Glasgow.
Killed in Action 16 September 1943.

SERVICE	PROMOTIONS	AWARDS
Joined RAF 1941		
Edinburgh 22 May 1941		
No. 1 ACRC 21 Jul 1941		
No. 1 ITW 16 Aug 1941		
Sth Africa 10 Nov 1941		
No. 45 Air Sch 10 Jan 1942		
No. 43 Air Sch 26 Apl 1942		
Presmoor 19 May 1942		
No. 3 PRC 29 Jun 1942		
No. 9 (O) AFU 26 Jul 1941		
No. 19 OTU 31 Aug 1942		
No. 1660 CU 5 Jan 1943		
617 Sqdn 10 Apl 1943		

Sergeant Daniel Allatson – front gunner – see Ken Brown's crew list.

Sergeant Austin Ainsworth Williams (1539602) Rear Gunner:
b 1919 Granetown, York.
Killed in Action 16 September 1943.

Joined RAF 1941
No. 3 RC 29 Jul 1941
No. 10 (S) RC 24 Nov 1941
RDU 16 Feb 1942
No. 10 AOS 6 Mar 1942
No. 11 STT 8 Jul 1942
No. 4 AGS 19 Sep 1942
No. 1660 CU 9 Nov 1942
207 Sqdn 17 Feb 1943
97 Sqdn 15 Mar 1943
617 Sqdn 25 Mar 1943

Flying Officer James Alexander Rodger (1070106) Navigator.
Trained with Wilson's crew, but did not fly on the raid.
b 1911 Edinburgh.
Killed in Action 16 September 1943.

Joined RAF 1940	
No. 3 RC 21 Sep 1940	No. 121558
No. 1 RW 15 Mar 1941	
No. 11 ITW 5 Apl 1941	
PAA 1 Aug 1941	
PTC 14 Aug 1941	
Miami 24 Aug 1941	
No. 31 PD 6 Dec 1941	
UK 26 Dec 1941	
No. 3 PRC 26 Dec 1941	
No. 5 AOS 16 Jan 1942	Plt Off 19 May 1942
617 Sqdn 1943	Fg Off 19 Nov 1942

Sergeant D.M. Buntaine
Front Gunner in Ken Brown's crew – replaced due to illness for the raid.
Died 1981.

SERVICE	PROMOTIONS	AWARDS
617 Sqdn Mar 1943		
No. 26 OTU 26 Mar 1944	F/Sgt	
	Plt Off 3 Mar 1945	
	Fg Off 3 Sep 1945	
Released 1946		

Sources

617 Squadron Personnel Interviewed or corresponded with
Ken Brown CGM; Edna Bark (Mrs); John Bryden; Tony Burcher DFM; Sidney Hobday DFC; Lance Howard DFC; Bill Howarth DFM; Jim Heveron; M.J. Hull (Mrs); George Johnson DFM; F. Payne; George Powell; Dick Roberts; Dave Rodger DFC; Keith Stretch; Len Sumpter DFC, DFM; F. Tees; Bill Townsend CGM, DFM; F. Waterman; Canon J.A. Williams DFC; Jim Watson DFC; Jim Clay DFC.

Next of Kin
Mrs G. Bramber (neé Maltby); Mrs B. Bjelland (neé Taerum); Mrs M. Bulloch (neé Gregory); Mrs L. Burpee; Glen Byers; Mrs R. Anderson; Mrs M.A. Gowrie; Mary Hopkins; Mrs R.A. Johnson; Mrs N. Knight; Mrs Molly Kirby (neé Franklin); Charles Marriott; Miss E.M. Nugent; Sidney Pulford; Mr S.A. Taylor; Richard Thrasher QC; Marner Young (neé Hopgood)

Marshal of the Royal Air Force Sir Arthur T. Harris Bt, GCB, OBE, AFC, LLD; Air Chief Marshal the Hon Sir Ralph A. Cochrane GCB, KCB, AFC; Air Commodore John Searby DSO, DFC; Group Captain Hamish Mahaddie DSO, DFC, AFC; Air Chief Marshal Sir Christopher Foxley-Norris GCB, DSO; Group Captain D.E. Verity; Group Captain F. Winterbotham CBE; Wing Commander Rupert Oakley DSO, DFC, AFC, DFM; Ralph Barker; Chaz Bowyer; Alfred Price; Martin Middlebrook; W.G. Ramsey (Editor *After the Battle* Magazine)

Newspapers
The Times; The Daily Telegraph; Daily Express; The Sunday Times; Sunday Express; Lincolnshire Echo; South Wales Argus (Ann Jones)

Magazines
The Aeroplane; Flight International (Stephen Birch); *Air Mail* (RAF Association)

Other Sources consulted
Ministry of Defence, (RAF Personnel Management Centre); Ministry of Defence, (AR8/AR9 Mrs Elderfield); Ministry of Defence, (Marine) Holland; Ministry of Defence, New Zealand; Ministry of Defence, Australia; Air Historical Branch; Canadian Photographic Unit; Bomber Command Association; Ray Callow; Bill Rust; Harry Pitcher DFM; Air Gunners Association; F.N. Gill; Aircrew Association; Danny Boon; RAFA

Helston – Bert Cawls; RAF Museum – J.M. Bruce; Australian War Memorial Museum; Bomber Command Association (NZ Section); 617 Association (Canadian Division); 617 Squadron RAF Scampton; Christ's Hospital; The Royal Society; Imperial War Museum: Department of Photographs (Ted Hine): Department of Film (Queenie Turner); The Commonwealth War Graves Commission (Mrs Wendy Wilcox, Mrs D. Brown)

Stadt Hamm; Stadt Arnsberg; Westfalisher Anzeiger; Stadt Archivund Emmerich

Geoff Richmond; John F. Grime DFC; M.R. Sargent; Jonathan Dimbleby; David Dimbleby; Stuart Stephenson (Lincolnshire Lancaster Committee); Alec Kennedy – Corpus Christi College, Cambridge; Dr E.V. Bevan – Trinity College, Cambridge; Roy Chappell (author of *Wellington Wings*); Jim Halliday; T.D. Taylor DFC; E.L. Penning; Ivor Stanbrook MP; Miss A. Berry; Squadron Leader C.H. Wood; Eric N. Rhodes; A.C. Nutter; M.J. Pitcher; Dr A.R. Collins MBE; Jeremy Hall; Jack Foreman; George Griffiths DFM; Don Dearn; John Harding; Edward A. McFadden; R.R. Chapman; Peter Sharpe; Dr Steve Ross; George Brown; S.R. White; Peter Moran; Professor J.E. Morpurgo BA; L.M. Baldwin MM; B.B. Halpenny; Stanley Richmond; Rev. F.M. Griffiths BSc; Sheila Ruskell; John Frost; Joe Corrigan; P.L. Rickson (Hawker Siddeley); Brian Oliver DFC (ex-106 Sqdn); George Brookes; J.G. de Conninck (British Aerospace); Rolls Royce (Michael Evans); Frank Waldron; The late Sir Barnes Wallis CBE; Wing Commander Wally Dunn OBE; Bob Gardiner; Hildegard Gaze – for the greatest help in translation work; Horst Muller whose help and support since I started my project, also enabled me to visit the dams in May 1980; Herman B. Van den Dool; Norbert Kruger; D.S. Drijver; A.A.F.M. van Riel; H.B. van Helden; Karl Schütte; Dr. Albert Speer; Arthur Askey; Dr. Fletcher; Rhodes House; Diana Birch; Des Richards; Jim Boys.

Public Records Office
References

Air 14-840	Air 22-8	Air 24-205
Air 14-2215	Air 22-78	Air 20-4821
Air 14-2220	Air 22-1663	Air 50-272
Air 14-2088	Air 8-1234	Air 27-2128
Air 14-2144	Air 20-4369	Air 2 Code 30
Air 14-844	Air 19-383	Air 30
Air 2-5944	Air 24-255	Avia 10
Air 2-2574	Air 29-276	

London Gazette – Honours and Awards.

Glossary

AAS. – Advanced Air School.
AC – Air Commodore.
ACDW – Air Crew Disposal Wing.
ACRC – Air Crew Recruit Centre.
ADC – Aide de Campe.
AFU – Advanced Flight Unit.

AGS – Air Gunnery School
AM – Air Marshal.
ANS – Air Navigation School.
AOC – Air Officer commanding.
AOS – Air Observer School.
AVM – Air Vice-Marshal.
B & GS – Bombing and Gunnery School.
CB – Commander of the Bath.
CBE – Commander of the British Empire.
C in C – Commander in Chief.
CF – Conversion Flight.
CGM – Conspicuous Gallantry Medal.
CPL – Corporal.
CTS – Combat Training School.
CU – Conversion Unit.
DFC – Distinguished Flying Cross.
DFM – Distinguished Flying Medal.
DSO – Distinguished Service Order.
EAOS – Elementary Air Observer School.
EFTS – Elementary Flying Training School
EOWS – Elementary Observer Wireless School.
FIS – Fighter Intercept School.
FLT/LT – Flight Lieutenant.
F/O – Flying Officer.
F/S – Flight Sergeant.
GBE – Knight Grand Cross of the British Empire.
G/C – Group Captain.
IAAS – Inland Area Aircraft School.
ITS – Initial Training School.

ITW – Initial Training Wing.
KBE – Knight Commander of the British Empire.
KCB – Knight Commander of the Bath.
LAC – Leading Aircraftman.
LFS – Lancaster Finishing School.
MBE – Member of the Order of the British Empire.
MID – Mentioned in Despatches.
MU – Maintainance Unit.
NCO – Non-commissioned Officer.
OAFU – Officer Advanced Flying Unit.
OBE – Officer of the Order of the British Empire.
OBS – Observer School.
OC – Officer Commanding.
OCU – Officers Cadet Unit.
OTU – Operational Training Unit.
OTW – Operational Training Wing.
PDC – Personnel Despatch Centre.
P/O – Pilot Officer.
RC – Recruit Centre.
RD – Recruit Depot.
RDU – Radio Direction Unit.
RTP – Recruit Training Pool
RW – Recruit Wing.
SAN – School of Navigation.
SFT – School of Flying Training.
SFTS – Services Flying Training School.
SGT – Sergeant.
S/L – Squadron Leader.
SRC – Signals Recruit Centre.
STT or S of TT – School of *Torpedo* Training.
TD – Training Depot.
TU – Training Unit.
WAAF – Womens Auxiliary Air Force.
W/C Wing Commander.
W/S – Wireless School
W & SS – Wireless and Signal School.
WU – Wireless Unit
VC – Victoria Cross.

Index